Updates in Advances in Lung Cancer

Volume Editor   *J.H. Schiller,* Madison, Wisc.

9 figures and 47 tables, 1997

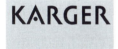

KARGER   Basel · Freiburg · Paris · London · New York ·
New Delhi · Bangkok · Singapore · Tokyo · Sydney

....

Joan H. Schiller

University of Wisconsin Comprehensive Cancer Center
Medical Oncology Section, Madison, WI 53792

Bibliographic Indices. This publication is listed in bibliographic services, including Current Contents® and Index Medicus.

© Copyright 1997 by S. Karger AG, P.O. Box, CH–4009 Basel (Switzerland)
Printed in Switzerland on acid-free paper by Reinhardt Druck, Basel
ISBN 3–8055–6557–7

....
Updates in Advances in Lung Cancer

Progress in Respiratory Research

Vol. 29

Series Editor   *C.T. Bolliger*, Basel

   Basel · Freiburg · Paris · London · New York ·
New Delhi · Bangkok · Singapore · Tokyo · Sydney

Contents

Preface

A short time ago, the previous volume of this book series (No. 28), entitled 'The Tobacco Epidemic', was published. It was the first volume under my editorship, which I had taken over from Prof. H. Herzog, Editor Emeritus. Among many other topics, the ill effects of voluntary and involuntary smoking were addressed in that book. Many thousands of original articles on the deleterious effects of tobacco consumption have been published. The most important damage is seen in the lungs where smoking is responsible for the development of chronic obstructive pulmonary disease and for up to 85–90% of lung cancers. The current global epidemic of cigarette smoking has led to a dramatic increase of this neoplasm, especially in women. Despite a lot of research, little has changed in the grim prognosis of lung cancer. Any new therapy represents thus a gleam of hope which is usually anxiously awaited.

When we heard about Dr. Joan H. Schiller's intention to publish her 'Updates in Advances in Lung Cancer', we invited her to contribute to our book series, making it a logical extension of the previous topic. It gives me great pleasure therefore to announce the publication of Dr. Schiller's 'Updates in Advances in Lung Cancer' as volume 29 in our series *Progress in Respiratory Research.* The chapter contents are well suited for this book series, the emphasis of which is on recent developments in pulmonary medicine. I am confident that this volume will generate a lot of interest in a large multidisciplinary readership.

Basel, September 1997

C.T. Bolliger
Series Editor

Foreword

Lung cancer remains a major cause of cancer deaths throughout the world. In many countries, it causes more cancer-related deaths than breast cancer, colon cancer, and prostate cancer combined. It is an extremely lethal neoplasm; 80–90% of patients who develop lung cancer will die of the disease.

The purpose of this book is to update cancer specialists, pulmonary physicians, and general practitioners about the many recent advances that have been made in the treatment, prevention, and biology of this disease. Despite the overall poor prognosis of lung cancer patients, a number of new chemotherapy drugs and regimens have recently been developed which result in a clinically meaningful improvement in survival for patients with non-small cell lung carcinoma; four chapters in this book are devoted to reviewing these new drugs and drug combinations. The activity of some of these newer chemotherapeutic agents as first- and second-line agents in small cell lung cancer is detailed. Advances in the management of locally advanced non-small cell and small cell lung cancer are reviewed, including combined-modality treatments involving chemotherapy, surgery, and radiation therapy. Results of recent trials involving chemoprevention agents for the prevention of lung cancer are reported. Finally, the many recent advances that have been made in understanding the molecular biology of this disease, and how they might impact on the clinical management of lung cancer patients, are discussed.

Joan H. Schiller

Schiller JH (ed): Updates in Advances in Lung Cancer. Prog Respir Res.
Basel, Karger, 1997, vol 29, pp 1–20

Chapter 1
··························

Lung Cancer Chemoprevention and Management of Carcinoma in situ

Daniel D. Karp [a], *Amy Law* [b]

[a] Cancer Clinical Research Office, Beth Israel Deaconess Medical Center, and
[b] Hematology/Oncology Division, The New England Medical Center, Boston,
Mass., USA

Introduction and Current Perspective

According to American Cancer Society estimates, in 1997 there will be 178,100 cases of lung cancer in the US with approximately 160,400 deaths [1], a 1% increase over 1996. At the highest level of government, tobacco is now recognized as an addictive drug warranting regulation by the Federal Drug Administration (FDA) [2]. Currently, approximately 46.3 million individuals (26% of US adults) smoke and an additional 40–50 million are former smokers [3]. While approximately 70% of smokers state they want to quit, only 2.5% succeed. Despite programs such as the Community Intervention Trial for Smoking Cessation (COMMIT), addiction to tobacco remains a major public health problem [4]. Long-term survivors of lung cancer are now conservatively estimated to have a risk of developing second primary cancers in 20% of cases and former smokers now comprise an increasing percentage of newly diagnosed cases of lung cancer [3]. Elimination of tobacco use as primary prevention of lung cancer remains a crucial intervention for teenagers and young adults. However, even if all persons stop using tobacco, the legacy of decades of smoking and other environmental and occupational risk factors in older adults will continue to produce huge numbers of new cases of lung cancer [5]. Furthermore, the risk of second primary tumors for people cured of primary cancers of the lung or head and neck remains high and does not seem to be effectively improved by smoking cessation alone [6]. Consequently, new prevention strategies must be developed [7]. The state-to-state variations in US cancer death rates suggest that changes in behavior and environmental exposure could

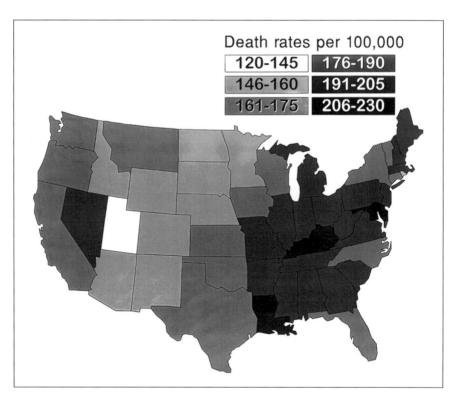

Fig. 1. Cancer death rates per 100,000 (from American Cancer Society 1997).

favorably affect the number of people dying from cancer (fig. 1). Utah has the lowest death rate (126/100,000) from all forms of cancer; the midwest farming states are next with rates ranging from 145 to 160, whereas the eastern industrial states have consistently higher death rates from cancer.

Ever since the landmark report by Hong et al [8] in 1990 that isotretinoin (13-*cis*-retinoic acid) could prevent second primary tumors (SPTs) in patients with previously treated squamous cell carcinoma of the head and neck, it has been hoped that similar strategies could reduce the incidence of lung cancer in current smokers, former smokers, and those with a previously treated aero-digestive tract malignancy. In that regard, SPTs must be distinguished from a recurrence of a previously treated cancer. Hong et al [8] defined SPTs as those cancers separated from the original tumor by 2 cm of normal epithelium or consisting of a different histologic type or recurring after an interval of greater than 5 years from the original cancer. Metachronous tumors are those that occur greater than 6 months from the original tumor whereas synchronous cancers are those diagnosed within 6 months of the original cancer. These distinctions are pivotal in chemoprevention research. SPTs occur with relatively constant frequency (i.e., 3% per year in a high-risk cohort). SPTs often result from a field defect and are usually not treatment related. It is the purpose of this article to focus on recent developments and current chemoprevention strategies to lend a framework for a practical approach to lung cancer chemoprevention and the treatment of in situ carcinoma.

As far back as 1922, Mori [9] reported that a diet low in fat-soluble vitamin A resulted in squamous metaplasia in the trachea and larynx of experimental animals. This observation and others stimulated the use of vitamins and other dietary supplements which now make up a billion dollar health food industry supported by millions of people seeking to prevent not only cancer, but the aging process in general Slaughter [10, 11], beginning in 1944 introduced the concept of multiplicity of origin of cancer – the so-called 'field cancerization' concept. In 1953, Slaughter et al [12] reported that 88 of 783 (11.2%) patients with oral cancer had two or more independent squamous cell carcinomas. They also found that abnormal and hyperplastic, often atypical epithelium, was found to surround virtually all oral cancers for varying distances. In 1957, Auerbach et al [13] reported epithelial changes in the lungs of 41 patients with bronchogenic cancer not present in the lungs of nonsmokers: hyperplasia, loss of cilia, and atypical cells. The authors defined carcinoma in situ as the presence of all three of these abnormalities. These two landmark studies demonstrated that, rather than being an isolated event, lung cancer is the end result of a series of genetic defects which accumulate in the bronchial tree following long-term exposure to tobacco and other carcinogens. Sporn et al [14] coined the term 'chemoprevention' in the mid-1970s, thereby defining an entire new area of cancer research: chemoprevention is the attempt to 'arrest or reverse pre-malignant cells during their progression to invasive malignancy, using physiologic mechanisms that are not toxic'. They further expanded this concept to develop interventions to retard the promotional events that make up the dynamic multistep process of carcinogenesis.

Multistep Carcinogenesis

It is now widely accepted that there is a sequence of oncogene activation events required for the development of an invasive lung cancer cell. This mechanism has been worked out in detail for colon cancer by Vogelstein et al [15]. Although the process of oncogene activation in bronchogenic cancer is less well understood, activation of six families of oncogenes – ras, raf, jun, fur, neu and myc – have been associated with lung cancer [16–19]. K-ras mutations occur in approximately 30% of adenocarcinomas in smokers [20]. Almost all the mutations occur in codon 12, usually with a G:T transversion. These mutations seem to be irreversible and remain silent until additional genetic alterations occur which promote further neoplastic development. K-ras mutations are rarely, if ever, detected in small cell cancer. Myc family DNA amplification and expression occurs frequently in small cell cancer as well as in approximately 10% of non-small cell lung cancer [21]. Furthermore, myc protein has been detected in 10/15 (67%) cases of surgically resected cancer in a study by Morkve et al [22]. It appears that abnormalities in p53 and genetic mutations of k-ras occur relatively early in the development of lung cancer [23].

Molecular Epidemiology: Finding Subjects at Increased Risk for Cancer

The causal pathway initiated by smoking includes a long list of factors: duration of smoking, depth of inhalation, chronic obstructive lung disease, and deposition of particulate matter. A second branch of the pathway includes chronic inflammation, release of proteolytic enzymes, release of oxygen radicals, and macrophage secretion of polypeptide growth factors [24]. Genetic susceptibility makes up the third component [25]. Knudson [26], in 1985, hypothesized that certain individuals have increased susceptibility to development of cancer based on the loss of a single gene which then makes it more likely that they could suffer a 'second hit' as a result of carcinogen exposure or some other mutational stimulus. This hypothesis connects sporadic and familial cancers and has a broad range of applicability for both basic and clinical research. Because only 10–15% of smokers ultimately develop lung cancer and it is they who could most benefit from chemoprevention [27], a variety of tests and models have been investigated to predict those at greatest risk for the development of lung cancer. Recent studies by Krontiris et al [28] have reported an association of variability in the HRAS1 minisatellite locus and the risk of several epithelial cancers, including lung cancer. A current active Eastern Cooperative Oncology Group (ECOG) study is expanding this observation by testing for the presence of such genetic polymorphisms in siblings with cancer to better define the concept of 'cancer families'. For many years, it has been held that individuals whose P_{450} cytochrome system metabolized the drug debrisoquine slowly were at higher risk of development of lung cancer [29], although a recent report by Shaw et al [30] questions that finding. A case-control study by Wu et al [31] reported increased bleomycin-induced chromosome damage in the lymphocytes of lung cancer patients. Spitz

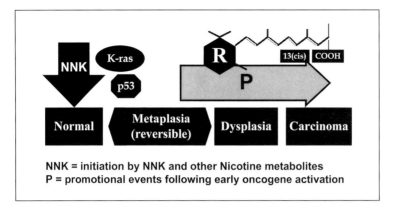

Fig. 2. Chemoprevention: using retinoids and other agents to attempt to slow the promotional events that lead from dysplasia to carcinoma. NNK = Initiation by NNK and other nicotine metabolites; P = promotional events following early oncogene activation.

et al. [32] have shown that patients with curatively treated upper aerodigestive cancer whose lymphocytes had increased sensitivity to mutagens had a higher risk of developing a second cancer. The recent identification of susceptibility genes for breast cancer (BRCA1, BRCA2) has further intensified interest in the search for new tests to identify individuals at highest risk for lung cancer who might benefit from specific intervention efforts.

Specific Chemopreventive Agents

For a drug to be effective as a chemopreventive agent, it should retard or reverse the long process of carcinogenesis: the progression from initiated or pre-malignant cells to severe dysplasia to the disease we call cancer [33] (fig. 2).

Vitamin A and Dietary Intake

A statistical decrease in the risk of lung cancer has been associated with increased consumption of green or yellow vegetables or carotenoids in 14 of 15 retrospective diet studies. Conversely, decreased serum retinol levels have been correlated with an increased risk of cancer [34]. There is still no compelling evidence to date, however, that a diet high in vitamin A will prevent lung cancer, although one study using retinyl palmitate has been suggestive of benefit (see Pastorino et al. [62] below).

The Retinoids

The retinoids are derivatives of vitamin A that are critical for epithelial differentiation, normal vision, and reproduction. Natural and synthetic reti-

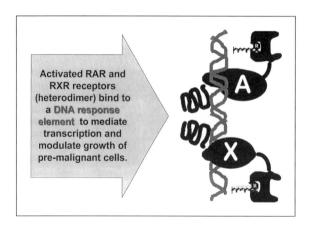

Fig. 3. Activated receptors bind as heterodimers to DNA response elements.

noids given at pharmacological doses can restore regulation of differentiation in vitro and so they were a natural choice for early studies [35, 36]. Retinoids are potent regulators of gene expression [37, 38] which work through an elaborate system which consists of a family of cytoplasmic retinoic acid binding proteins (CRABP) as well as intranuclear retinoic acid receptors [39]. The family of retinoid nuclear receptors consists of two main types: RAR and RXR – each with at least three subtypes named α, β, γ. RAR-α RAR-β RAR-γ have been localized to chromosomes 17q21.1, 3p24, and 12q13, respectively whereas RXR-α RXR-β RXR-γ have been localized to chromosomes 9q34.3, 6p21.3, and 1q22–23 [23]. The nuclear retinoid receptors are homologous to the steroid/thyroid hormone family [40]. Structurally, RARs have three separate domains: a transcription activating domain, a DNA binding domain, and a hormone binding domain together with a hinge region. Once an appropriate retinoid reaches the cytoplasm, it enters or is transported to the nucleus where it binds to the hormone binding domain. The RARs and RXRs then bind to DNA as heterodimers and inhibit protein kinase C, modulate growth of premalignant cells, and suppress evolution of premalignancy to overt neoplasia. In particular, abnormality or loss of detectable RARβ has been associated with lung cancer and lung cancer cell lines [41, 42] and a majority of oral premalignant lesions responding clinically to 13cRA treatment have been shown to have an increase in the RAR-β level [43] (fig. 3). The retinoid receptors do not appear to bind directly to 4-HPR (all-trans-4-hydroxyphenyl retinamide) and it is therefore termed a 'pro-drug' whereas 9-cis and other retinoids bind to both RAR and RXR [44] (fig. 4). The startling discovery

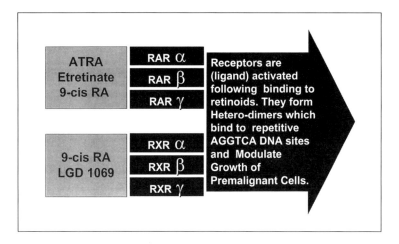

Fig. 4. Retinoic acid receptor binding and mechanism of action.

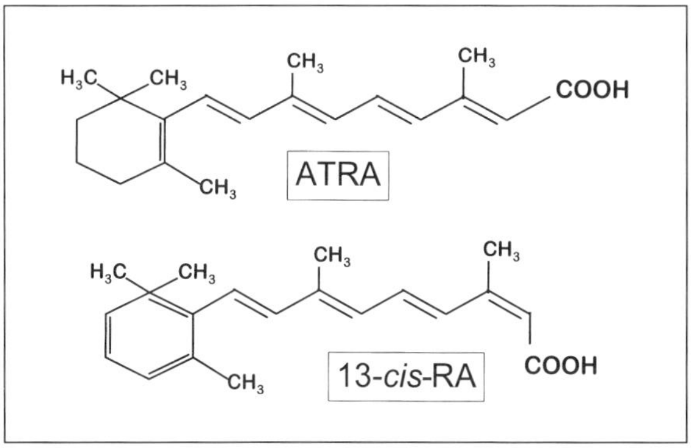

ATRA

13-*cis*-RA

Fig. 5. All-*trans*-retinoic acid (ATRA) and isotretinoin (13-*cis*-RA) are both in clinical use.

that ATRA can induce differentiation and clinical remission in patients with acute promyelocytic leukemia was a major milestone in the quest for a more biologically based approach to the prevention and treatment of malignant disease [45] (fig. 5).

Etretinate is a second-generation compound, and the arotinoids, such as LGD 1069 and tazarotene, are third-generation agents which have been modified in the ring structure to provide more differentiation and gene regulation activity [44] (fig. 6).

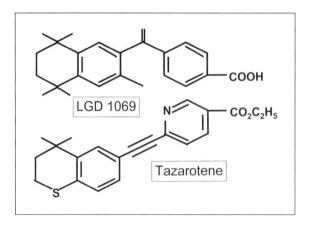

Fig. 6. LGD 1069 and tazarotene are third-generation arotinoids with multi-ring structure.

Carotenoids

β-carotene: Carotene is a dimer of the active moiety retinol which is plentiful in green and yellow vegetables. It has appeared to have an advantage over preformed vitamin A and was a popular agent for clinical trials over the past decade. However, recent reports from three large studies have not shown benefits in chemoprevention of lung cancer and two have even suggested that overall death rate and deaths from lung cancer are *higher* in smokers who take pharmacological doses of β-carotene. The Alpha-Tocopherol Beta-Carotene (ATBC) Study [46] was carried out on 29,133 male Finnish smokers. Those participants who took 20 mg of β-carotene showed 18% more lung cancers and 8% more deaths than those ATBC participants who did not. The Physician's Health Study began in 1982 and ended on December 31, 1995. It was a randomized 2×2 factorial trial that accrued 22,070 US male physicians aged 40–84 years. Eleven percent of the participants were smokers and 51% had smoked at some time in their life when recruited to the study. Participants took either β-carotene 50 mg every other day, or a placebo and 325 mg of aspirin or a placebo on alternate days. This study showed no benefit or harm from β-carotene regarding cancer or cardiovascular events (table 1). The Beta-Carotene and Retinol Efficacy Trial (CARET) study involved 18,314 smokers, former smokers or aerospace workers exposed to asbestos. It was terminated in January 1996 after a planned interim analysis showed a 28% higher rate of lung cancer and 17% overall death rate in those participants taking β-carotene. The participants were notified to stop taking their study medications and obtain follow-up and a physical exam. The dose of β-carotene in that

Table 1. Placebo-controlled studies have shown no β-carotene benefit for smokers

Study	Size	Drugs	Result
ATBC Study NEJM 1994	29,133 male Finnish smokers	β-Carotene 20 mg and α-tocopherol	Higher rates of lung CA and deaths 18% and 8%
CARET β-Carotene and retinol evaluation trial	18,314 14,254 smokers, and 4,060 asbestos workers	β-Carotene 30 mg and vitamin E 25,000 IU	Higher rates of lung CA (388) and deaths (974) 28% and 17%
Physicians' Health Study 1982 C. Hennekens	22,071 males 11% current and 51% former smokers; age 40–84	β-Carotene 50 mg qod and aspirin 325 mg	44% decrease in myocardial infarct (ASA)
Women's Health Study J. Buring, Sc.D.	40,000 professional women aged ≥45	Vitamin E 600 IU and aspirin 100 mg	Ongoing (β-Carotene stopped)

Table 2. The Carotene and Retinol Efficacy Trial (CARET) was halted by NCI on January 11, 1996

Double-blind placebo-controlled study
β-Carotene 30 mg and vitamin E 25,000 IU
18,314 participants (56% M/44% F)
   14,254 with smoking history
      9,408 smokers
      4,846 former smokers (34%)
   4,060 asbestos workers
Events: 388 cancers, 974 deaths
   28% more lung cancer/17% more deaths

study was 30 mg/day, which is equivalent to 50,000 IU of vitamin A or about five medium-sized carrots (table 2). Current trials using β-carotene have been halted. The ongoing Women's Health Study is continuing using vitamin E and aspirin only. Although it is disappointing that β-carotene demonstrated no benefits for those participants, these important studies have shown the unquestionable importance of placebo-controlled studies in chemoprevention research. There are many individuals who have been taking β-carotene on the premise that it could not be harmful. The ATBC and CARET studies suggest the opposite. Smokers *should not take β-carotene* supplements. They should stop smoking.

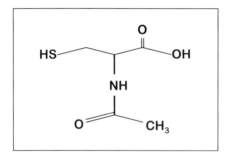

Fig. 7. N-acetylcysteine.

N-Acetylcysteine (NAC)

Cigarette smoke contains dozens of carcinogens: benzene, benzo(a)pyrene, NNK (4-(methylnitrosamino)-1-(3-pyridyl)-1butanone), crotonaldehyde, epoxides, free radicals, and peroxides. It has been estimated that each puff of smoke contains 10^{16} oxidant molecules, a level which can overwhelm the glutathione detoxification system (GSH) [47]. Of note, oxidant exposure can rapidly activate k-ras oncogene DNA [48]. NAC is an aminothiol precursor of reduced GSH that has been used since the early 1960s as a mucolytic agent [49]. When taken orally, NAC is rapidly absorbed, deacetylated, and incorporated into the intra- and extracellular GSH stores [50]. It is highly effective in ameliorating acetaminophen hepatotoxicity and is safe when given in doses of 30 g daily for several days [51]. NAC detoxifies reactive electrophiles and free radicals through reduction of oxygen species and by conversion to cysteine which supports GSH biosynthesis and acts directly as an antioxidant or as substrate in the GSH redox cycle. In vitro, NAC decreases carcinogen-induced DNA damage, inhibits type IV collagenases as well as chemotaxis, invasion, and metastasis by some malignant cells [52]. The European Organization for Research and Treatment of Cancer (EORTC) is studying NAC at a dose of 600 mg daily as well as vitamin A in the form of retinyl palmitate (300,000 IU) in a 2×2 factorial designed study – EUROSCAN. Side effects of NAC have been very low with 86% of participants taking NAC alone reporting essentially no toxicity [53] (fig. 7).

Other Agents

Vitamin E (α-tocopherol) is a naturally occurring antioxidant. Epidemiologic and dietary studies have suggested that there is an inverse relationship between vitamin E intake and the incidence of lung cancer [54]. Data by Dimery et al [55] suggested that the apparent reduction in toxicity of high-dose 13-*cis*-retinoic acid by a tocopherol makes this an attractive combination for future study (fig. 8).

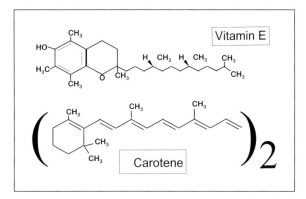

Fig. 8. Vitamin E (α-tocopherol) and β-carotene.

Selenium: This essential trace element was first associated with cancer protection in the 1960s [56]. Epidemiologic evidence suggests that selenium plays a major role in protecting against the development of human cancer [57]. In addition, there is a variability in the environmental selenium levels across the country. A recent study by Clark et al [58] suggested that selenium supplementation (200 µg/day) in the form of brewer's yeast tablet is associated with a reduction in the risk of lung cancer. The study was a randomized controlled trial on a total of 1,312 patients. The primary endpoint was designed to evaluate the effect of selenium supplementation on the risk of developing new basal and squamous carcinomas of the skin. No protection was seen for skin cancer. The protocol was modified in 1990 to analyze total incidence and mortality of cancers of the lung, prostate, colon and rectum. There is a statistically significant 39% reduction in incidence (77 cancers in the selenium group and 119 in controls) and 48% reduction in mortality (29 deaths in the selenium group and 57 deaths in controls). However, other randomized prevention trials have not shown a similar degree of protection against cancer. Whether this inconsistency is the result of the dose of selenium, the form of selenium, or specific cancers studied is unclear [59]. Further randomized studies are needed to confirm the results before any public health recommendations regarding selenium supplementation can be made.

Isothiocyanates such as phenethyl isothiocyanate (PEITC) and benzyl isothiocyanate (BITC) occur in conjugated form in a wide variety of cruciferous vegetables and are released in the course of chewing by the enzyme myrosinase. PEITC and BITC inhibit lung cancer production by NNK and benzo(a)pyrene respectively. They decrease DNA adducts and increase urinary excretion of detoxified NNK [60]. They must be present at the time of carcinogen exposure to be effective, however, and thus would be most likely to confer benefit when incorporated into a daily nutrition program that begins early in life.

Chalcones are synthetic open chain flavanoids. They are stilbene derivatives which bind to steroid receptors and have some structural similarity to tamoxifen. Another agent, *myo*-inositol, is a nontoxic phytate derivative that has been used clinically to treat diabetic neuropathy. Both chalcones and *myo*-inositol inhibit NNK pulmonary adenoma formation in female A/J mice [58]. These and other novel inhibitors of pulmonary carcinogenesis hold promise for new interventions in prevention of lung cancer [61].

Preventing Second Primary Tumors in Patients at Risk

At the present time, there is no standard chemoprevention therapy nor is there an FDA-approved agent for the prevention of lung cancer or development of SPTs. In the original high-dose placebo-controlled 13-*cis*-retinoic acid study of Hong et al [8] in head and neck cancer patients with approximately 50 participants in each arm, there were three lung cancers that developed in the control group whereas none developed in the treatment group. Pastorino et al [62] recently reported a study of 307 patients with resected stage I non-small cell lung cancer who were randomized to receive 300,000 units of retinol palmitate daily or observation only for 12 months. With a median follow-up of 46 months, 51 SPTs developed in 47 patients. There were 18 SPTs in the treatment arm and 29 in the control arm. The overall estimated survival rate at 5 years was 62 versus 54% (p = 0.44). The current Intergroup Study 91025 is randomizing patients with stage I (T1N0M0 or T2N0M0) resected non-small cell lung cancer to 13-*cis*-retinoic acid (30 mg daily) versus placebo. This study, which began in September 1992, is being coordinated by the M.D. Anderson Cancer Center and includes participation by the Cancer and Acute Leukemia Group B (CALGB), community CCOPs, ECOG, the North Central Group (NCCTG), and the Southwest Group (SWOG). Approximately 1,400 patients will be required for completion of this truly national protocol which is due to complete accrual on time in April 1997. At the annual review in March 1996, 844 of 938 participants (90%) had successfully completed the run-in phase of the protocol that precedes the randomization to 13-*cis*-retinoic acid or placebo. Only 1 case of grade 4 toxicity has been recorded. One hundred seven cases have reached the study endpoints. There have been 28 new primaries and 66 relapses. Already, valuable lessons have been learned from this study. The involvement of a group of active and committed thoracic surgeons as well as medical and radiation oncologists has been crucial to the successful accrual enjoyed by this ambitious study. As a result, patients have been highly motivated to enter and maintain a high rate of compliance. Yearly endpoint analysis has been consistent with the Lung Tumor Cooperative Group findings of 2–3% new SPTs per year in such a group [63]. However, the actual number of recurrences and SPTs in such a large carefully staged patient group will be a valuable by-product of this study. The EORTC Chemoprevention Study in Lung Cancer (EUROSCAN, EORTC No. 08871), began in March 1988 and completed accrual in January 1994. Patients with curative treatment for T1, 2, or 3, N0 squamous cancer of the oral cavity and larynx, or resected non-small cell lung cancer, were randomized in a factorial design to receive retinol palmitate orally (150,000–300,000 IU/day for 12–24 months) and/or NAC 600 mg/day orally for 24 months. As of the January 1996 analysis, with data on 2,440 participants (1,449 head & neck, 991 lung), there have been 204 recurrences in the head & neck patients (14.1%) and 266 recurrences in the lung patients (26.8%). There have been a total of 133 SPTs (5.5%) with 83 in the head & neck group (5.7%) and 50 in the lung group (5.0%) [64].

Intermediate Endpoints

There are multiple intermediate endpoints which have been employed in chemoprevention research to identify useful new strategies. These include oncogene activation, presence of micronuclei, assays of DNA ploidy, biochemical measures such as levels of ornithine decarboxylase, prostaglandin synthetase, and blood group antigens. Exfoliated cells can now be analyzed for the presence of bronchial metaplasia. However, changes in such biomarkers have not been validated with being associated with decreased risk of developing lung cancer. Nevertheless, these important surrogates are of key importance in providing information rapidly and allowing studies with reduced sample size [65].

Clinical Trials of Chemopreventive Agents

Lee et al [66] recently reported the results of a prospective trial of the use of isotretinoin in 86 heavy smokers. The authors were attempting to confirm the results of an earlier uncontrolled French trial which reported that etretinate (25 mg/day for 6 months) could produce a decline in extent of squamous metaplasia in the bronchial epithelium of smokers [67]. Participants were taken from an original group of 152 smokers who were found on bronchoscopy to have frank dysplasia or metaplasia in greater than 15% of the biopsy sections examined (metaplasia index). They then received either isotretinoin (1.0 mg/kg) or placebo for 6 months and were re-evaluated carefully with bronchoscopic biopsies taken from six sites. Interestingly, a large number of participants in both groups stopped smoking during the study and the metaplasia index decreased over time in 54% of the treatment group and 59% of the control group. Although this was termed a 'negative' study, it did document that a placebo-controlled study with bronchoscopically determined intermediate endpoint is feasible in lung cancer chemoprevention trials and smoking cessation can be associated with a sizable decrease in the degree of bronchial metaplasia over a 6-month period. It is also quite possible that markers other than metaplasia index will emerge which are more useful in such studies.

Other Studies

The NCI Early Detection Branch is conducting a 16-year, 74,000 participant randomized screening trial for prostate, lung, colorectal, and ovarian cancer – the PLCO trial – to assess whether studies such as flexible sigmoidoscopy, chest x-ray, digital rectal exam, serum PSA, CA-125 and transvaginal ultrasound can reduce cancer mortality in men and women aged 60–74 [68]. ECOG is evaluating immunostaining of induced sputum specimens, shed antigens, and the presence of growth factor elevation in bronchial lavage fluid for the early detection of second primary tumors in patients with curatively resected stage I non-small cell lung cancer. Hopefully, these and other studies will yield a wealth of important information for the screening and prevention of lung cancer.

Carcinoma in Situ

Carcinoma in situ (T_{is}) is also known as stage 0 non-small cell lung cancer and denotes microscopic squamous cell carcinoma, usually discovered during the course of bronchoscopy. The in situ designation refers to squamous cell type only and ordinarily does not apply to the other histologic forms of lung cancer. Occult lung cancer (Tx), on the other hand, denotes, a patient whose sputum or bronchial washings contain malignant cells although tumor is not visualized on imaging studies or by bronchoscopy [69]. The usual scenario consists of a smoker with a negative chest x-ray and no other evidence of disease who is being evaluated for a productive cough or hemoptysis. Histologically, in situ squamous carcinoma ranges from thickening, erythema, and loss of the normal bronchial mucosal longitudinal ridges to ulceration and full-thickness atypia of squamous epithelium [70]. The multistage model assumes that atypical squamous metaplasia and dysplasia precede the development of in situ carcinoma, which in turn evolves to invasive cancer [71]. Frost et al [72] showed that cancer developed in approximately 12% of individuals with moderate atypia and 37% of those with severe atypia. Therefore, careful pathologic evaluation is the key to proper treatment of these lesions.

Surgery

T_{is} and occult lung cancers are noninvasive by definition and, *if localized*, should be curable with surgical resection. A very useful practice guideline algorithm has recently been published by Cameron et al [73]. Any patient with a positive sputum cytology should have a full medical history and complete physical exam, including a careful otolaryngology evaluation to rule out the possibility of a lesion in the oral cavity, larynx, pharynx or hypopharynx. Additional studies include a chest x-ray, serum chemistries, a CT scan of the chest and upper abdomen and pulmonary function tests. Selective bronchoscopy with brushings and washings of the individual lobes should be the next step. If a localized unequivocally malignant area is identified, lobectomy or pneumonectomy would then be a curative approach to the problem. If any question remains regarding the degree of dysplasia, a waiting period followed by repeat bronchoscopy may be a valuable maneuver; during that period, the patient should absolutely refrain from smoking. In the event that no localizing cytology is obtained, the patient should stop smoking and have careful follow-up with history and physical exam and repeat bronchoscopy. A 6- to 12-week interval is a reasonable standard. If a specific lesion is identified later, surgery can still hopefully provide long-term successful treatment.

Photodynamic therapy (PDT) is a research-based technique first developed in the 1980s and has the ability to detect clinically 'early' lung cancers. This technique takes advantage of the increased autofluorescence of malignant cells either alone or following use of photosensitizers such as hematoporphyrin derivatives [74]. PDT appears to be an alternative to surgical resection in some patients with localized bronchogenic carcinoma [75]. A bronchoscopically fitted laser light source can be used to visualize and ablate localized tumor [76]. This technique has the advantage of preserving lung tissue in a population of patients who often have compromised pulmonary function and are at high risk for multiple cancers. However, such an approach requires close follow-up and if PDT is unsuccessful, patients will require surgical 'salvage' and continued surveillance for second primary cancers.

Chemopreventive Therapy

Patients with T_{is} have probably accumulated a sequence of oncogene activation events and other mutational stimuli. Therefore, a single agent such as vitamin-based therapy may not be adequate. It probably requires a more aggressive combination agent approach which inhibits different phases of the carcinogenesis process. This is also true of long-term survivors of small cell cancer, who have an alarming rate of second primary non-small cell cancers. Plans are underway to initiate such a study [R. Wihn, pers. commun.] in small cell cancer survivors using combination therapy with agents such as retinoids, vitamin E, and NAC.

Conclusions and Considerations for the Future

Chemoprevention is a new discipline and yet has made important strides over the last decade. A coalition of patients, health professionals, researchers, and governmental staff now exists to promote increased research and intervention programs nationally. Several large studies have recently finished accrual and are likely to provide new valuable insights soon. It is imperative that practicing oncologists seek out and support programs for their high-risk patients. β-Carotene has been shown in two large studies to increase the overall death rate in active smokers as well as increase the risk of developing cancer. The mechanism is unknown at this time. This experience with β-carotene has stimulated multiple researchers to investigate this paradox and has taught us that uncontrolled use of health food supplements should be avoided in smokers until it is clear that a particular agent is beneficial Health professionals must increase efforts to eliminate tobacco use in our citizens – especially the young ones who can benefit most. Smoking cessation alone appears to be capable of reversing metaplasia in some individuals. Studies with retinoids and retinyl palmitate have suggested benefits in preventing second primary cancers. Patients with a long history of smoking, strong family history and/or extensive occupational or environmental exposures should be urged to take advantage of chemoprevention protocols when available. In the meantime, a diet containing proper amounts of cruciferous vegetables and fruits is an important health practice that should be encouraged early in life since many dietary components must be present at the time of carcinogen exposure to confer benefit. An impressive catalog of intermediate markers and activated oncogenes is now available to provide surrogate endpoints for new research. New techniques such as phototherapy may soon add a new dimension to the diagnosis and conservative therapy for very early stage and in situ lung cancer. Such advances, while important, are very costly. Patient care costs have proven to be a substantial barrier to the performance of chemoprevention studies in some settings. Nurses, dedicated support staff, and data managers are necessary to ensure a high level of patient compliance and data accuracy in chemoprevention programs. Physicians, health plan administrators and payers, as well as representatives of the pharmaceutical industry and governmental agencies must form partnerships to take chemoprevention to the next level of success. As our country approached the 20th century, lung cancer was a rare disease. As we now approach the 21st century, we have an increasing number of powerful tools to diagnose, prevent, and treat lung cancer. If we employ the tools at our disposal, we can hope to make a great impact on this lethal 100-year epidemic of lung cancer and benefit our patients in a meaningful way.

References

1   CA Cancer J Clin 1997;47:5–27.
2   President's message to teenage smokers. The Boston Herald, Thurs Aug 22, 1996, p 1.
3   Strauss G, DeCamp M, Dibiccaro E, Richars W, Harpoie D, Healey E, Sugarbaker D: Lung cancer diagnosis is being made with increasing frequency in former cigarette smokers! Proc Am Soc Clin Oncol 1995;14:362.
4   Major professional organizations join forces in fight against tobacco addiction. Reuters Health Information News Service, June 25, 1996.
5   Garfinkel L, Stellman SD: Smoking and lung cancer in women: Findings in a prospective study. Cancer Res 1988;48:6951–6955.
6   Karp DD, Vaughan CW, Willet B, Heeren T, DiMartino NA, Vincent ME, Picardi ME, Hong WK: Long-term survival following induction chemotherapy for squamous cell carcinoma of the head and neck: A ten-year follow-up; in Salmon SE (ed): Adjuvant Therapy of Cancer. V. Orlando, Grune & Stratton, 1987, pp 119–128.
7   Brown CC, Kessler LG: Projections of lung cancer mortality in the United States, 1985–2025. J Natl Cancer Inst 1988;80:43–51.
8   Hong WK, Lippman SM, Itri LM, Karp DD, Lee JS, Bjers RM, Schantz SP, Kramer AM, Lotan R, Peters L: Prevention of second primary tumors with isotretinoin in squamous-cell carcinoma of the head and neck. N Engl J Med 1990;323:795–801.
9   Mori S: The changes in the para-ocular glands which follow the administration of diet low in fat soluble A; with note of the effects of the same diets on the salivary glands and the mucose of the larynx and trache. Johns Hopkins Hosp Bull 1922;33:357–359.
10 Slaughter DP: The multiplicity of origin of malignant tumors; Collective review. Int Abstr Surg 1944;79:89–98.
11 Slaughter DP: Multicentric origin of intraoral carcinoma. Surgery 1946;20:133–146.
12 Slaughter DP, Southwick HW, Smejkal W: 'Field cancerization' in oral stratified squamous epithelium. Clinical implications of multicentric origin. Cancer 1953;6:963–968.
13 Auerbach O, Patrick TG, Stout AP, et al: The anatomical approach to the study of smoking and bronchogenic carcinoma: A preliminary report of forty-one cases. Cancer 1956;9:76–83.
14 Sporn MB, Dunlop NM, Newton DL, Smith JM: Prevention of chemical carcinogenesis by vitamin A and its synthetic analogs (retinoids). Fed Proc 1976;35:1332–1338.
15 Vogelstein B, Fearon ER, Hamilton SR, Kern SE, Preisinger AC, Leppert M, Nakamura V, White R, Smits A, Bos JL: Genetic alterations during colorectal tumor development. N Engl J Med 1988; 319:525–535.
16 Huber MH, Lee JS, Hong WK: Chemoprevention of lung cancer. Semin Oncol 1993;20:128–141.
17 Vogt Pk, Bos TJ: Jun: Oncogenes and transcription factors. Adv Cancer Res 1990;55:1.
18 Angel P, Karin M: The role of Jun, Fos, and the AP-1 complex in cell proliferation and transformation. Biochim Biophys Acta 1991;1072:129.
19 Distel RJ, Spiegelman BM: Protooncogene c-fos as a transcription factor. Adv Cancer Res 1990; 55:37.
20 Westra WH, Slebos RJC, Offerhaus GJA, Goodman SN, Evers SG, Kensler TW, Askin FB, Rodenhurs S, Hruban RH: K-ras oncogene activation in lung adenocarcinomas from former smokers: Evidence that k-ras mutations are an early and irreversible event in the development of adenocarcinoma of the lung. Cancer 1993;72:432.

21   Johnson BE: The role of myc, jun, and fos Oncogenes in human lung cancer; in Pass HI, Mitchell JB, Johnson DH, Turrisi AT (eds): Lung Cancer: Principles and Practice. Philadelphia, Lippincott-Raven, 1996, pp 83–98.

22   Morkve O, Halvorsen OJ, Stangeland L, et al: Quantitation of biological markers (p53, c-myc, Ki-67, and DNA ploidy) by multiparameter flow cytometry in non-small cell lung cancer. Int J Cancer 1992;52:851.

23   Hong WK, Lippman SM, Hittelman WN, et al: Retinoid chemoprevention of aerodigestive cancer: From basic research to the clinic. Clin Cancer Res 1995;1:677–686.

24   Islam SS, Schottenfeld D: Declining FEV_1 and chronic productive cough in cigarette smokers: A 25-year prospective study of lung cancer incidence in Tecumseh. Michigan. Cancer Epidemiol Biomarkers Prev 1994;3:289.

25   Schottenfeld D: Epidemiology of lung cancer; in Pass HI, Mitchell TB, Johnson DH, Turrisi AT (eds): Lung Cancer: Principles and Practice. Philadelphia, Lippincott-Raven, 1996, pp 305–321.

26   Knudson AG: Hereditary cancer, oncogenes, and anti-oncogenes. Cancer Res 1985;45:1437.

27   Mattson ME, Pollack ES, Cullen JW: What are the odds that smoking will kill you? Am J Public Health 1987;77:425–431.

28   Krontiris TG, Devlin B, Karp DD, et al: An association between the risk of cancer and mutations in the HRAS1 minisatellite locus. N Engl J Med 1993;329:517–523.

29   Speirs CJ, Murray S, Davies DS, et al: Debrisoquine oxidation phenotype and susceptibility to lung cancer. Br J Clin Pharmacol 1990;29:101–109.

30   Shaw GL, Falk RT, Tucker MA, et al: Debrisoquine metabolism and lung cancer risk. Proc Annu Meet Am Assoc Cancer Res 1994;35:1753A.

31   Wu X, Hsu TC, Annegers JF, Amos CI, Fueger JJ, Spitz MR: A case-control study of nonrandom distribution of bleomycin-induced chromatid breaks in lymphocytes of lung cancer cases. Cancer Res 1995;55:557–561.

32   Spitz MR, Hoque A, Trizna Z, Schantz SP, Amos CI, King TM, Bondy ML, Hong WK, Hsu TC: Mutagen sensitivity as a risk factor for second malignant tumors following malignancies of the upper aerodigestive tract. J Natl Cancer Inst 1994;86:1681–1684.

33   Sporn MB: Carcinogenesis and cancer – Different perspectives on the same disease. Cancer Res 1991;51:6215–6218.

34   Willett WM, McMahon B: Diet and cancer: An overview. N Engl J Med 1984;310:633.

35   Smith MA, Parkinson DR, Cheson BD, et al: Retinoids in cancer therapy. J Clin Oncol 1992;10: 839–864.

36   Sporn MB, Dunlop NM, Newton DL, Smith JB: Prevention of chemical carcinogenesis by vitamin A and its synthetic analogs (retinoids). Fed Proc 1976;35:1332.

37   Borden EC, Lotan R, Levens D, et al: Differentiation therapy of cancer: Laboratory and clinical investigations (meeting report). Cancer Res 1993;53:4109–4115.

38   Gudas L, Hu L: The regulation and gene expression by retinoids in normal and tumorigenic epithelial cells. Proc Annu Meet Am Assoc Cancer Res 1993;34:588–589.

39   Kitamura M, Shirasawa T, Mitarai T, et al: A retinoid-responsive cytokine gene. MK, is preferentially expressed in the proximal tubules of the kidney and human tumor cell lines. Am J Pathol 1993; 142:425–431.

40   Wagner H, Ruckdeschel JC: Screening, early detection, and early intervention strategies for lung cancer. Cancer control. J Moffitt Cancer Center 1995;2:493–502.

41   Gebert JF, Moghal N, Frangioni JV, Sugarbaker DJ, Neel BG: High frequency of retinoic acid receptor beta abnormalities in human lung cancer. Oncogene 1991;6:1859–1868.

42   Lotan R, Sozzi G, Ro J, Lee JS, Pastorino U, Pilotti S, Kurie J, Hong WK, Xu XC: Selective suppression of retinoic acid receptor b expression in squamous metaplasia and in non-small cell lung cancers compared to normal bronchial epithelium. Proc Am Soc Clin Oncol 1995;14:165.

43   Lotan R, Xu XC, Lippman SM, Ro JY, Lee JS, Lee JJ, Hong WK: Suppression of retinoic acid receptor in premalignant oral lesions and its upregulation by isotretinoin. N Engl J Med 1995;332: 1405–1410.

44   Lippman SM, Heyman RA, Kurie JM, Benner SE, Hong WK: Retinoids and chemoprevention: Clinical and basic studies. J Cell Biochem 1995;(suppl 22):1–10.

45 Huang M, Ye Y, Chen S, et al: Use of all-*trans*-retinoic acid in the treatment of acute promyelocytic leukemia. Blood 1988;72:567–572.

46 The Alpha-Tocopherol, Beta-Carotene Cancer Prevention Study Group: The effect of vitamin E and beta-carotene on the incidence of lung cancer and other cancers in male smokers. N Engl J Med 1994;330:1029–1035.

47 Van Zandwijk N: N-acetylcysteine and glutathione: Antioxidant and chemopreventive properties, with special reference to lung cancer. J Cell Biochem 1995;(suppl 22):24–32.

48 Cochrane CG: Cellular injury by oxidants. Am J Med 1991;91:23–30.

49 Webb WR: Clinical evaluation of a new mucolytic agent acetyl cysteine. J Thorac Cardiovasc Surg 1962;44:330–343.

50 Meister A, Anderson M: Glutathione. Annu Rev Biochem 1983;52:711–760.

51 Miller LF, Rumack BH: Clinical safety of high oral doses of acetylcysteine. Semin Oncol 1983; 10(suppl 1):76–85.

52 De Flora S, Cesarone CF, Balansky RM, Albini A, D'Agostini F, Bennicelli C, Bagnasco M, Camoirano A, Scatolini L, Rovida A, Izzotti A: Chemopreventive properties and mechanisms of N-acetylcysteine. The experimental background. J Cell Biochem 1995;(suppl 22):33–41.

53 Van Zandwijk N, Pastorino U, de Vries N, Dalesio O: EUROSCAN: The European Organization for Research and Treatment of Cancer (EORTC): Chemoprevention study in lung cancer. Lung Cancer 1993;9:351–356.

54 Menkes MS, Constock GW, Vuilleumier JP, et al: Serum beta-carotene, vitamin A and E, and selenium and the risk of lung cancer. N Engl J Med 1986;315:1250.

55 Dimery I, Shirinian M, Heyne K, et al: Reduction in toxicity of high-dose 13-*cis*-retinoic acid with a tocopherol (abstract). Proc Am Soc Clin Oncol 1992;11:399.

56 Shamberger RJ, Frost DV: Possible protective effect of selenium against human cancer. Can Med Assoc J 1969;100:682.

57 Van den Brandt PA, Goldbohm RA, van't Veer P, Bode P, Dorant E, Hermus RJJ, Sturmans F: A prospective cohort study on selenium status and the risk of lung cancer. Cancer Res 1993;53:4860–4865.

58 Clark LC, Combs GF, Turnbull BW, et al, for the Nutritional Prevention of Cancer Study Group: Effect of selenium supplementation for cancer prevention with carcinoma of the skin: A randomized controlled trial JAMA 1996;276:1957–1963.

59 Blott WJ, Li JY, Taylor PR, et al: Nutrition intervention trials in Linxian, China: Supplementation with specific vitamin/mineral combinations, cancer incidence, and disease-specific mortality in the general population. J Natl Cancer Inst 1993;85:1483–1492.

60 Estenser RD, Wattenberg LW: Studies of the chemopreventive effects of myo-inositol on benzo[a]pyrene-induced neoplasia of the lung and forestomach of female A/J mice. Carcinogenesis 1993;14: 1975–1977.

61 Wattenberg L: Chalcones, *myo*-inositol and other novel inhibitors of pulmonary carcinogenesis. J Cell Biochem 1995;(suppl 22):162–168.

62 Pastorino U, Infante M, Maioli M, et al: Adjuvant treatment of stage I lung cancer with high-dose vitamin A. J Clin Oncol 1993;11:1216–1222.

63 Thomas P, Rubinstein L, and the Lung Cancer Study Group: Cancer recurrence after resection: T1N0 non-small cell lung cancer. Ann Thorac Surg 1990;49:242.

64 Kirkpatrick A: EORTC Central Office. Personal communication.

65 Huber MH, Lippman SM: Chemoprevention strategies; in Pass HI, Mitchell JB, Johnson DH, Turrisi AT (eds): Lung Cancer: Principles and Practice. Philadelphia, Lippincott-Raven, 1996, pp 351 358.

66 Lee JS, Lippman SM, Benner SE, et al: Randomized placebo-controlled trial of isotretinoin in chemoprevention of bronchial squamous metaplasia. J Clin Oncol 1994;12:937–945.

67 Misset JL, Mathe G, Santelli G, Gouveia J, Homasson JP, Sudre MC, Gaget H: Regression of bronchial epidermoid metaplasia in heavy smokers with etretinate treatment. Cancer Detect Prev 1986;9:167–170.

68 Current Clinical Trials: Oncology. National Cancer Institute PDQ 1996;3:199.

69 Beahrs OH, Henson DE, Hutter RVP, Kennedy BJ (eds): Handbook for Staging of Cancer. American Joint Committee on Cancer. Philadelphia, Lippincott, 1993, pp 129–136.

70   Travis WD, Linder J, Machay B: Classification, histology, cytology, and electron microscopy; in Pass HI, Mitchel JB, Johnson DH, Turrisi AT (eds): Lung Cancer: Principles and Practice. Philadelphia, Lippincott-Raven, 1996, pp 361–395.

71   Melamed MR, Zaman MB, Flehinger BH, Martini N: Radiologically occult in situ and incipient invasive epidermoid lung cancer: Detection by sputum cytology in a survey of asymptomatic cigarette smokers. Am J Surg Pathol 1997;1:5.

72   Frost JK, Ball WC Jr, Levin ML, Tockman MS, Erozan YS, Gupta PK, Eggleston JC, Pressman NJ, Donithan MP, Kimball AW: Sputum cytopathology: Use and potential in monitoring the workplace environment by screening for biological effects of exposure. J Occup Med 1986;28: 692–703.

73   Cameron R, Fringer J, Taylor C, Gilden R, Figlin RA: Practice guidelines for non-small cell lung cancer. Cancer J Sci Am 1996;2(suppl):69–77.

74   Hung J, Lam S, LeRiche C, et al: Autofluorescence of normal and malignant bronchial tissue. Laser Surg Med 1991;11:99.

75   Lam S, Palcic B, McLean D, et al: Detection of early lung cancer using low-dose Photofrin II. Chest 1990;97:333.

76   Edell ES, Corese DA: Photodynamic therapy in the management of early superficial squamous cell carcinoma as an alternative to surgical resection. Chest 1992;102:1319.

D.D. Karp, MD, Director, Cancer Clinical Research Office,
Beth Israel Deaconess Medical Center,
Kirstein Hall Room 158, 330 Brookline Avenue, Boston, MA 02215 (USA)
Tel. (617) 667 1910, Fax (617) 975 8030, E-Mail daniel.karp@bidmc.harvard.edu

Schiller JH (ed): Updates in Advances in Lung Cancer. Prog Respir Res.
Basel, Karger, 1997, vol 29, pp 21–34

Chapter 2

....

The Role of Chemotherapy and Surgery in the Treatment of Locally Advanced Nonsmall Cell Lung Cancer

David W. Johnstone, Hazem Y. Afifi, Richard H. Feins

Division of Cardiothoracic Surgery, University of Rochester Medical Center,
Rochester, N.Y., USA

Introduction

Lung cancer has become the leading cause of cancer death in American men and women. Nonsmall cell lung cancer (NSCLC) is diagnosed in 140,000 Americans each year, most of whom will die of their disease. Only 15–20% of these patients will present with resectable intraparenchymal disease. Up to 50% will have evidence of locally advanced tumors, by virtue of mediastinal adenopathy or direct invasion of adjacent structures. Successful treatment of patients with locally advanced nonsmall cell carcinoma remains elusive, despite advances in tumor staging, surgical technique, chemotherapeutic agents, and radiotherapy. Multimodality strategies are essential to achieve local and distant control over these cancers and have become the norm in current clinical research for locally advanced NSCLC.

Staging of NSCLC has become standardized under the current International Staging System (tables 1, 2) [1]. Lymph node metastases to hilar (N1), ipsilateral mediastinal (N2), or contralateral mediastinal/supraclavicular (N3) positions are associated with a poorer prognosis overall and a high incidence of recurrence following surgical resection alone. Similarly, tumors that extend to the chest wall (T3) or central mediastinal structures (T4) have a poorer prognosis than tumors of similar nodal status without direct tumor invasion. The majority of locally advanced tumors are thus stage IIIa-b. Tumors causing malignant pleural effusions (T4) are excluded from this discussion.

Table 1. New International Staging System: TNM Classification [1]

Primary tumor (T)

TX: Tumor proved by the presence of malignant cells in bronchopulmonary secretions but not visualized radiographically or bronchoscopically, or any tumor that cannot be assessed in a retreatment staging

T0:   No evidence of primary tumor

Tis: Carcinoma in situ

T1:   A tumor that is ≤ 3.0 cm in greatest dimension, surrounded by lung or visceral pleura, and without evidence of invasion proximal to a lobar bronchus at bronchoscopy

T2:   A tumor >3.0 cm in greatest diameter or a tumor of any size that either invades the visceral pleura or has associated atelectasis or obstructive pneumonitis extending to the hilar region. At bronchoscopy the proximal extent of demonstrable tumor must be within a lobar bronchus or at least 2.0 cm distal to the carina. Any associated atelectasis or obstructive pneumonitis must involve less than an entire lung

T3:   A tumor of any size with direct extension into the chest wall (including superior sulcus tumors), diaphragm, or the mediastinal pleura or pericardium without involving the heart, great vessels, trachea, esophagus, or vertebral body, or a tumor in the main bronchus within 2.0 cm of the carina without involving the carina

T4:   A tumor of any size with invasion of the mediastinum or involving the heart, great vessels, trachea, esophagus, vertebral body, or carina, or the presence of malignant pleural effusion

Nodal involvement (N)

N0:   No demonstrable metastasis to regional lymph nodes

N1:   Metastasis to lymph nodes in the peribronchial or the ipsilateral hilar region, or both, including direct extension

N2:   Metastasis to ipsilateral mediastinal lymph nodes and subcarinal lymph nodes

N3:   Metastasis to contralateral mediastinal lymph nodes, contralateral hilar lymph nodes, and scalene or supraclavicular lymph nodes

Distant metastasis (M)

M0:   No (known) distant metastasis

M1:   Distant metastasis present

Stage III NSCLC is comprised of a heterogenous group of tumors with different natural histories, prognoses, and treatment options. For example, T3N0 carcinomas with chest wall invasion can be resected en bloc anticipating a 45–50%-year survival with no adjuvant therapy. By contrast, tumors with N2 disease carry an overall 5-year survival of 15%; yet within this subgroup, tumors with single-station intracapsular nodal metastases appear to have a 5-year survival approaching 30% with complete resection alone [3–5].

Table 2. NSCLC Stage Grouping by TNM Classification

	New International Staging System [1]			AJCC Revised Stage Grouping [37]			
Stage 0	Tis	N0	M0				
Stage I	T1-2	N0	M0	IA	T1	N0	M0
				IB	T2	N0	M0
Stage II	T1-2	N1	M0	IIA	T1	N1	M0
				IIB	T2	N1	M0
					T3	*N0*	*M0*
Stage IIIA	T3	N0-2	M0	IIIA	T3	N1	M0
	T1-3	N2	M0		T1-3	N2	M0
Stage IIIB	Any T	N3	M0	IIIB	Any T	N3	M0
	T4	Any N	M0		T4	Any N	M0
Stage IV	Any T	Any N	M1	IV	Any T	Any N	M1

Trials Using Surgery Alone

Several large single-institution series of surgically resected stage III non-small cell carcinoma have demonstrated a survival benefit in selected patients with limited disease. These are all retrospective series culled from larger patient populations with locally advanced disease, yet they provide a basis of current multimodality strategies by defining those stage III patients with better overall prognosis who may benefit most from neoadjuvant therapy, as well as suggesting that surgery alone may have a role in the treatment of highly selected patients.

In a retrospective analysis of 151 patients undergoing complete resection of N2 lung cancer, Martini and Flehinger [5] reported an overall 5-year survival rate of 30%. Preoperative mediastinoscopy was not routinely used for lymph node staging. Most of these patients received postoperative radiation therapy. Survival was better in patients with clinical N0-1 disease, T1-2 tumors, and single-station nodal metastases. Patients with clinically unsuspected N2 disease had a significantly better 5-year survival than those with clinically apparent N2 disease (37–47 vs. 9%). Patients undergoing thoracotomy with incomplete resection or no resection were excluded from survival analysis.

In a series of 141 patients staged with mediastinoscopy, Pearson et al. [6] found significantly better 5-year survival in patients having a false-negative mediastinoscopy (41%) than patients with positive mediastinoscopy (15%).

Watanabe et al. [3] found similar results: clinical N2 patients undergoing complete resection had a 5-year survival rate of 20% compared with 33% in patients with completely resected but clinically unsuspected N2 disease. The overall survival of N2 patients undergoing resection in this series was only 17%.

Naruke et al. [7] emphasized the benefit of mediastinal lymph node dissection in a review of 426 patients undergoing resection for N2 disease. 242 patients with complete resections had a 5-year survival of 19.2%. 345 patients undergoing mediastinal node dissection had a 15.5% 5-year survival compared with 6.7% in patients who did not have a mediastinal dissection; the latter group had more advanced disease and worse performance status, so the survival benefit of mediastinal node dissection for patients with N2 disease remains unclear in this study.

Goldstraw et al. [4] recently reviewed 149 patients undergoing resection of unsuspected N2 disease. These patients were staged preoperatively with CT scanning and/or mediastinoscopy. Despite rigorous preoperative investigation of patients with potentially resectable NSCLC, mediastinal node dissection revealed nodal metastases in 26% of the patients. A complete resection was performed in 85% of this group and was associated with an actuarial 5-year survival of 20.1%.

The role of surgery in locally invasive NSCLC involving the mediastinum was examined by Martini et al. [8] in a retrospective analysis of 102 patients with T3-4 N0-1 tumors treated by surgical resection. Fifty-five percent had incomplete resections and most of these received intraoperative brachytherapy to control residual tumor. With complete resection the 5-year survival was 30% versus 14% with incomplete resection. Eighty-one percent of postoperative survivors had persistent or recurrent disease. The authors concluded that patients with mediastinal invasion, unlike with T3 tumors involving the peripheral chest wall, have a poor survival following surgery alone, and should be considered for neoadjuvant therapy.

These clinical studies provide the basis of current understanding of the therapeutic benefits and limits of surgical resection in locally advanced NSCLC. Though the data from these studies is retrospective with some methodological flaws, there appears to be a survival benefit from complete resection in patients with limited N2 disease (single station or clinically unsuspected). This group is a small subset of patients with stage III NSCLC that can be difficult to identify prospectively. Given the 20–30% false-negative rate of current CT scanning for detecting nodal metastases, routine use of mediastinoscopy is necessary to identify these patients with more accuracy – a practice that has little support among thoracic surgeons at this time. As Shields [9] has observed, when the inoperable, the unresectable, the incompletely resected, and the recurrences following apparent complete resection are stripped away, a tiny proportion of patients with node-positive stage III NSCLC realize long-term benefit from surgical resection.

Combined Modality Therapy

In an effort to improve outcomes for most patients with stage III tumors, combined modality strategies have been employed. Combined modality trials can be divided into those with surgical resection and those without it. Surgical adjuvant neoadjuvant protocols have been carried out over the past 15 years and are gradually reshaping the approach to this difficult subset of lung cancer patients.

Surgical trials can be grouped into those using preoperative chemotherapy or chemoradiotherapy (neoadjuvant or induction therapy), and those using postoperative chemo- or chemoradiotherapy only (adjuvant therapy). Preoperative chemotherapy alone permits greater dose intensity without incurring the risk of associated radiation toxicity, and allows higher radiation doses to be given postoperatively. Combined preoperative chemotherapy and radiation may produce more response in the primary tumor mass and thus make a complete resection more feasible, while providing satisfactory control of micrometastatic disease. The rationale for postoperative chemotherapy is to control unrecognized micrometastatic disease, since most recurrences are distant.

Adjuvant Chemotherapy Trials

Early adjuvant chemotherapy trials demonstrated no significant benefit. The Veterans Administration Lung Cancer Study Group compared surgery alone versus surgery followed by chemotherapy using either methotrexate and nitrogen mustard or methotrexate and cyclophosphamide [10]. The 5-year survival rates were 26 and 25% in the adjuvant chemotherapy and surgery groups, respectively. A similar later trail failed to show any benefit of adjuvant chemotherapy [11].

The introduction of cisplatin-based chemotherapy and improved staging resulted in controlled, randomized studies which attempted to address the role of chemotherapy in stage II and III NSCLC. The Lung Cancer Study Group (LCSG) completed a number of trials in this area.

In LCSG 772, 141 patients with completely resected stages II (T2N1) or III (any T3 or any N2) adenocarcinoma or large cell carcinoma were randomized to immunotherapy (intrapleural BCG/levamisole) or CAP chemotherapy (cyclophosphamide, doxorubicin, and cisplatin) every 4 weeks for six courses. Complete resection was defined as negative surgical margins and absence of tumor in the highest mediastinal node resected. The average dose of CAP chemotherapy received was 58% of the full protocol dosage. Distant recurrences predominated in both groups. A median survival advantage of 7 months with CAP was demonstrated (22.5 vs. 15.5 months), the median time to recurrence was increased by 7 months, but overall survival was not significantly improved [12].

In LCSG 791, adjuvant chemotherapy was evaluated in 172 patients with incompletely resected stage II and III NSCLC of all histologies. Incompletely resected disease was defined as the presence of positive surgical margins or evidence of tumor in the highest paratracheal node sampled during mediastinal dissection [13]. All patients received radiation (4,000 cGy in a split course), and the treatment group also received CAP. Compliance with chemotherapy was poor: only 51% of the patients completed all six cycles. There was no difference in local recurrence rates but overall freedom from recurrence at 5 years favored the group receiving chemotherapy and radiation (27 vs. 13%) [14]. For the CAP/RT patients, failure-free survival (14 months) and median survival (20 months) were prolonged by 6–7 months compared with patients receiving radiation alone. The survival benefit seen in this study is roughly equivalent to the time spent receiving adjuvant therapy.

In LCSG 853, 188 patients with completely resected stage II and III NCSLC were randomized to immediate combination chemotherapy (CAP) versus delayed chemotherapy administered at the time of the first systemic relapse. Local recurrences were treated with radiation. Median time to recurrence (19.5 months) and median survival (32.7 months) did not differ between groups. This study was unable to demonstrate either a recurrence-free survival benefit or an overall survival benefit for patients receiving immediate chemotherapy following complete resection of stage II-III NSCLC [15].

From Japan, The Study Group of Adjuvant Chemotherapy for Lung Cancer reported on 333 patients of stages I-III NSCLC, randomized to observation versus six cycles of adjuvant cisplatin, doxorubicin, and uracil-UFT (tegafur). Although there was no overall survival advantage, when the groups were separated by lymph node status, a survival advantage was noted for those receiving adjuvant chemotherapy [16].

Trials of adjuvant chemotherapy in NSCLC have not demonstrated a clear benefit. Though distant recurrence rates appear lower, this has not translated into a survival advantage. One reason for this is the use of relatively ineffective systemic therapy. There has been debate whether CAP is inferior to the cisplatin/vindesine combinations used more recently, but the latter regimen has not produced spectacularly better results. New, more effective agents are needed, and recently introduced drugs offer some promise in this regard. A second reason for the failure of adjuvant trials may be the difficulty in achieving target dosages in this population. Finally, the LSCG trials may have sought unrealistic median survival benefits for the regimens used and the trial sizes [16].

The recently completed US Intergroup Trial, INT-0115, may address some of these issues. This trial randomized patients with completely resected stage II and IIIa (N1-2) disease to thoracic radiation (50.4 Gy) or concurrent cisplatin/ etoposide and thoracic radiation. Accrual was complete in late 1996 with over 400 patients randomized. Follow-up and preliminary analysis are pending.

The observation that most relapses in early stage disease occur in distant sites, despite apparent complete surgical resection, has generated interest in adding systemic therapy to the management of earlier stage NSCLC. To evaluate the role of adjuvant chemotherapy in poorer prognosis early stage disease, the National Cancer Institute of Canada Clinical Trials Group has activated a phase III randomized study (NCIC BR.10) of adjuvant chemotherapy with vinorelbine and cisplatin in completely resected NSCLC (T2N0 or T1-2N1).

Neoadjuvant Chemotherapy Trials

Preoperative chemotherapy or chemoradiotherapy (induction or neo adjurant therapy) has several theoretical foundations: (1) cytoreduction of the local tumor and nodal metastases may increase the chance of complete resection and reduce intraoperative tumor dissemination; (2) early systemic therapy may eradicate micromatastases; (3) an intact tumor blood supply permits more effective delivery of chemotherapeutic agents; (4) therapy given prior to resection permits an estimate of tumor response.

Phase II trials have shown the feasibility of preceding surgical resection with induction chemotherapy in patients with stage III NSCLC [17–20]. The Memorial Sloan-Kettering Cancer Center reported a 7-year series of 136 patients with histologically confirmed stage IIIa NSCLC treated with preoperative MVP chemotherapy (mitamycin, vinblastine, and cisplatin). Those who responded underwent thoractomy and, if persistent nodal disease was encountered, received supplemental postoperative radiation. All patients received two postoperative courses of chemotherapy. Of 136 patients treated, the major response rate was 77%, the complete resection rate was 65%, and the complete pathologic response rate was 21%. Treatment-related mortality was 5%. The overall 5-year survival was 17% with a median survival of 19 months [2].

In a phase II trial from the Toronto group, 39 patients with histologically proven stage IIIa N2 disease were treated with preoperative MVP chemotherapy, followed by resection if a response was seen, and two postoperative courses of MVP. The overall response rate was 64% and the complete resection rate was 46% (18/39). Treatment-related mortality was 15% (6/39) due to sepsis and bronchopleural fistulae. Median survival was 18.6 months with a 3-year survival of 40% [21].

The LCSG (LCSG 881) randomized 67 patients with technically unresectable, histologically proven stage III disease (N2 or T4) to receive either preoperative MVP chemotherapy or preoperative radiation (44 Gy). Patients demonstrating partial or complete responses were eligible for resection. Of 57 evaluable patients, the overall response rate was 46%, the complete resection rate was 36% (18/57), and the treatment-related mortality was 12.5%. Definitive survival data was not available from this study, and the trial design was not intended to demonstrate a difference between the treatment arms [22].

The Cancer and Leukemia Group B (CALGB 8935) conducted a multi-institutional phase II trial, in which 74 patients with histologically proven stage IIIa N2 disease received induction cisplatin/vinblastine. Patients who did not progress underwent resection and then received two additional courses of chemotherapy and 59.4 Gy of radiation. Sixty-three patients (85%) underwent thoracotomy, and 23 (31%) had a complete resection with an operative mortality of 3.2%. Overall 3-year survival was 23% [23].

The Dana-Farber Cancer Institute series, reported by Elias et al. [24], included 54 patients with stage IIIA NSCLC who were treated with CAP neoadjuvant chemotherapy. The objective response rate was 39% with an 8% complete clinical response and a 44% complete resection rate. The 5-year survival was 20% with a median survival of 33.5 months for completely resected patients.

Two single-institution randomized trials of neoadjuvant chemotherapy have supported the findings of these phase II trials. Rosell [25] reported 60 patients with IIIA disease who were randomized to receive three preoperative courses of mitomycin, ifosfamide, cisplatin chemotherapy versus surgery alone. Both groups received postoperative radiation. In the group receiving preoperative chemotherapy, the response rate was 60% with an 83% complete resection rate. The neoadjuvant group had significantly better median survival (26 vs. 8 months) than the group treated with surgery alone. A trial from the M.D. Anderson Cancer Center yielded similar results. Sixty patients were randomized to receive three preoperative cycles of cyclophosphamide, etoposide, and cisplatin followed by surgery, versus surgery alone, in resectable stage IIIA NSCLC. The response rate to chemotherapy was 35%, and a substantial difference in median survival was found (64 vs. 11 months) [26]. These encouraging results are tempered by the small size of these trials, lack of mandated mediastinoscopy, and poor survival of the control arm patients.

A small, randomized trial from The National Cancer Institute has reported interim results. Twenty-seven patients with histologically proven N2 disease were randomized to perioperative cisplatin/etoposide chemotherapy (2 cycles preoperatively, 4 cycles postoperatively) versus surgery plus mediastinal radiation (54–60 Gy). There was a 62% partial response rate to chemotherapy and 85% in each group underwent resection. There was no significant difference between the treatment arms in overall survival or disease-free interval [27].

Neoadjuvant Chemoradiotherapy Trials

Several trials have examined the benefit of combined chemotherapy and radiation followed by surgical resection. The LCSG completed a neoadjuvant trial of cisplatin (75 mg/m^2), 5-FU, and partially concurrent radiation (30 Gy) in 85 patients with stage IIIa-b disease. The response rate was 56%, the resection rate was 52%, and the 3-year survival was 20% [28, 29].

The Cancer and Leukemia Group B conducted a phase II trial of concurrent chemotherapy and radiation in 41 patients with surgically staged IIIa disease [30]. Chemotherapy consisted of cisplatin (100 mg/m^2), vinblastine, and 5-FU, and was given concurrently with radiation (3,000 cGy in 15 fractions). A final cycle of chemotherapy and additional radiation (3,000 cGy) were given postoperatively. This study included some better prognosis stage IIIa patients (20% of the patients had T3N0-1 disease). The response rate was 51% and the complete resection rate was 61%. Treatment-related mortality was significant (15%). The median survival was 15.5 months, which is similar to survival rates seen with nonoperative strategies [31, 32].

The two largest phase II trials of concurrent preoperative chemoradiation demonstrate the feasibility of this strategy. The Rush-Presbyterian Study included 130 patients with clinically staged IIIa-IIIb disease who were treated with preoperative cisplatin (60 mg/m^2), 5-FU, +/– etoposide and simultaneous radiation (40 Gy split-course delivered over four cycles). Operative eligibility was determined before neoadjuvant therapy, and excluded patients with clinical N3 and extensive T4 disease. Eighty-five patients were eligible for resection, including 18 patients with T3N0 disease and 5 patients with T4 tumors. Of these, 62 (73%) underwent thoracotomy, and 58 (68%) had a complete resection. Operative mortality was 5%. Median survival for the 85 patients was 22 months, with an overall 3-year survival of 40% [33, 34].

The Southwestern Oncology Group (SWOG) enrolled 126 patients with pathologically staged IIIA-B disease to receive induction chemotherapy (cisplatin and etoposide) with concurrent radiotherapy (45 Gy continuous in 25 fractions). All patients underwent thoracotomy unless their disease progressed during induction treatment. Of the 126 patients, 107 (85%) were explored and 89 (71%) had a complete resection. Operative mortality was 6% and treatment-related mortality was 10%. Three-year survival was 26%. Interestingly, there was not significant survival difference between stages IIIa and IIIb. The strongest predictor of survival was the absence of tumor in mediastinal nodes at surgery [35, 36].

Issues for Trials

There is good evidence from these studies that neoadjuvant and adjuvant chemotherapy and radiation can be conducted with modest toxicity in patients with locally advanced NSCLC. It is also evident that surgical resection can be carried out in most patients following induction therapy with relatively low morbidity. It appears that survival is improved over series of surgical resection alone, although this difference is small and the patients subsets are not clearly comparable. The results of these trials must be interpreted with caution because of several common study flaws:

(1) Pretreatment staging has been neither uniform nor systematic. Few early studies performed mediastinoscopy in all patients, though this has now become more standardized with widespread acceptance of the current staging system and recognition of the prognostic importance of nodal metastases.

(2) The inclusion of better prognosis tumors (T3N0-1) or stage IIIB tumors (T4 or N3) makes it difficult to compare study results. It is essential that future studies mandate histologic staging and avoid admixture of patients from significantly different prognostic groups, so that survival benefits, if seen, can be correctly assigned to patients prospectively. A revision of the New International Staging System has been completed, which should help address this problem. Specifically, T3N0 tumors will move from stage IIIA to stage IIB (table 2) [37].

(3) Neoadjuvant regimens vary. Cisplatin-based chemotherapy predominates in these studies, but different doses and drug combinations have been used. Similarly, radiation has been delivered in split or continuous courses at different total dosages.

(4) Eligibility criteria for resection differ. Some trials mandated thoracotomy for all patients, some for those who showed no disease progression, while others offered surgery to responders only. It is difficult, therefore, to compare resectability rates and survival, and to assess the impact of chemotherapy.

(5) It is unclear whether the results seen in neoadjuvant trials can be translated to the larger community setting where close coordination between

oncologists and surgeons may be lacking, and surgical experience with difficult pulmonary resections may be limited. A multidisciplinary coordinated approach is essential to accurately stage these patients, inform them of the risks involved, and shepherd them through a prolonged period of intensive treatment. Pulmonary toxicity, seen particularly in trials using mitomycin [2, 22], has been less frequent with newer chemotherapeutic regimens, but careful attention to perioperative fluid management and bronchopulmonary hygeine is still essential to avoid major complications. Surgical challenges arise due to toxicity of preoperative chemotherapy and radiation, frequent hilar fibrosis and loss of tissue planes following induction therapy, and the need for complete resection and accurate nodal staging in the face of dense scar and/or residual tumor. The low operative mortality in these series reflects the experience and collaboration of thoracic oncologists working at major cancer centers or institutions with strong thoracic oncology programs.

(6) No large randomized trial of neoadjuvant therapy for locally advanced NSCLC has been completed. The trials described here have established a firm basis for such a study. A large prospective trial is now underway to determine whether neoadjuvant therapy with surgical resection is better than high-dose chemoradiation alone. The Thoracic Intergroup is conducting a randomized trial comparing neoadjuvant chemoradiation (cisplatin/etoposide $+45$ Gy delivered concurrently) plus surgery versus chemoradiation alone (cisplatin/etoposide $+61$ Gy). All patients receive two additional cycles of chemotherapy. This trial, INT-0139, requires pretreatment pathologic staging and excludes stage II and IIIB disease. Accrual to this trial from the major US cooperative oncology groups has been modest but steady.

Future Directions

Recent phase II and phase III trials have demonstrated the feasibility of combined modality treatment in stage III NSCLC, with modest improvements in survival. However, the majority of patients ultimately die of distant disease. Thus, the search for more effective systemic therapy remains paramount. Several new agents with novel mechanisms and toxicities are under investigation in clinical trials. Paclitaxel, an inhibitor of mitosis, Vinorelbine, a vinca alkaloid, and Gemcitabine, an antimetabolite, have attracted great interest [38]. The role of cytokines in the treatment of lung cancer is uncertain since initial results with interferons and interleukin-2 have been mixed [39, 40]. As the efficacy of these newer agents become well defined they will likely be incorporated into new neoadjuvant strategies.

As new agents and therapeutic strategies are developed for the treatment of NSCLC, the means to conduct large randomized phase III clinical trials within reasonable time frames will remain essential. The US Thoracic Intergroup was recently formed from the major cooperative oncology groups as a vehicle to enlarge the pool of institutions participating in clinical trials, speed accrual to large studies, and avoid competing protocols between cooperative groups. This is an important organizational step toward defining the role of multimodality therapy in thoracic malignancies, and will hopefully serve as a model for other fields of oncology in the future.

References

1 Mountain CA: New international staging system for lung cancer. Chest 1986;89(suppl):225–233.
2 Martini N, Kris MG, Flehinger BJ, Gralla RJ, Bains MS, Burt ME, Heelan R, McCormack PM, Pisters KM, Rigas JR, Rusch VW, Ginsberg RJ: Preoperative chemotherapy for stage IIIa (N2) lung cancer: The Sloan–Kettering experience with 136 patients. Ann Thorac Surg 1993;55: 1365.
3 Watanabe Y, Shimizu J, Oda M, Hayashi Y, Watanabe S, Tatsuwawa Y, Suzuki M, Takashima T: Aggressive surgical intervention for N2 non-small cell cancer of the lung. Ann Thorac Surg 1991; 51:253–261.
4 Goldstraw P, Mannam GC, Kaplan DK, Michail P, Shields TW: Surgical management of non-small cancer with ipsilateral mediastinal lymph node metastasis (N2 disease). J Thorac Cardiovasc Surg 1994;107:19–28.
5 Martini N, Flehiger BJ: The role of surgery in N2 lung cancer. Surg Clin North Am 1987;67: 1037–1049.
6 Pearson FG, Delarue NC, Ilves R, Todd TR, Cooper JD: Significance of positive superior mediastinal lymph nodes identified at mediastinoscopy in patients with resectable cancer of the lung. J Thorac Cardiovasc Surg 1982;83:1–11.
7 Naruke T, Goya T, Tsuchiya R, Suemasu K: The importance of surgery to non-small cell cancer of the lung with mediastinal lymph node metastasis. Ann Thorac Surg 1988;46:603–610.
8 Martini N, Yellin A, Ginsberg RJ, Bains MS, Burt ME, McCormack PM, Rusch VW: Management of non-small lung cancer with direct mediastinal involvement. Ann Thorac Surg 1994;58:1447–1451.
9 Shields TW: The significance of ipsilateral mediastinal lymph node metastasis (N2 disease) in non-small cell carcinoma of the lung. A commentary. J Thorac Cardiovasc Surg 1990;99:48–53.
10 Shields TW, Robinette CD, Keehn MS: Bronchial carcinoma treated with adjuvant cancer chemotherapy. Arch Surg 1974;90:329–333.
11 Shields TW, Higgins GA, Humphrey EW, Matthews MJ, Keehn RJ: Prolonged intermittent adjuvant chemotherapy with CCNU and hydroxyurea after resection of carcinoma of the lung. Cancer 1982; 50:1713–1721.
12 Holmes E: Surgical adjuvant therapy of non-small cell lung cancer. J Surg Oncol 1989;42(suppl 1): 26–33.
13 Lad T, Rubinstein L, Sadeghi A: The benefit of adjuvant treatment for resected locally advanced non-small cell lung cancer. J Clin Oncol 1988;6:9–17.
14 LCSG Statistical Report 1989.
15 Figlin RA, Piantodosi S: A phase III randomized trial of immediate combination chemotherapy vs. delayed combination chemotherapy in patients with completely resected stage II and III non-small cell carcinoma of the lung. Chest 1994;106(suppl):310–312.
16 Johnson DH: Adjuvant chemotherapy for non-small cell lung cancer. Chest 1994;106(suppl):313–317.

17 Rusch VW: Resection of stage III non-small cell lung cancer following induction therapy. World J Surg 1995;19:817–822.

18 Kirn DH, Lynch TJ, Mentzer SJ, Lee TH, Straus GM, Elias AD, Skarin AT, Sugarbaker DJ: Multimodality therapy of stage IIIa, N2 non-small cell lung cancer: Impact of preoperative chemotherapy on resectability and downstaging. J Thorac Cardiovasc Surg 1993;106:696.

19 Takita H, Regal AM, Anttwkowiak JG, Rao UNM, Botsoglou NK, Lane WW: Chemotherapy followed by lung resection in inoperable non-small cell lung carcinomas due to locally far-advanced disease. Cancer 1986;57:630.

20 Bitran JD, Golomb HM, Hoffman PC, Albain K, Evans R, Little AG, Purl S, Skosey C: Protochemotherapy in non-small cell lung carcinoma: An attempt to increase surgical resectability and survival. A preliminary report. Cancer 1986;57:44.

21 Burkes RL, Ginsberg RJ, Shepherd FA, Blackstein ME, Goldberg ME, Waters PF, Patterson GA, Todd T, Pearson FG, Cooper JD, Jones D, Loodwood G: Induction chemotherapy with mitomycin, vindesine, and cisplatin for stage III unresectable non-small lung cancer: Results of the Toronto Phase II Trial. J Clin Oncol 1992;10:580.

22 Wagner H, Lad T, Piantadosi S, Ruckdeschel JC: Randomized phase II evaluation of preoperative radiation therapy and preoperative chemotherapy with mitomycin, vinblastine, and cisplatin in patients with technically unresectable stage IIIa and IIIb non-small cell cancer of the lung. LCSG 881. Chest 1994;106(suppl):348.

23 Sugarbaker DJ, Herndon J, Kohman LJ, Krasna MJ, Green MR: The Cancer and Leukemia Group B Thoracic Surgery Group: Results of Cancer and Leukemia Group B Protocol 8935. A multi-institutional phase II trimodality trial for stage IIIa (N2) non-small cell lung cancer. J Thorac Cardiovasc Surg 1995;109:473.

24 Elias AD, Sakarin AT, Gonin R, Oliynk P, Stomper PC, O'Hara C, Socinski MA, Sheldon T, Maggs P, Frei E 3rd: Neoadjuvant treatment of stage IIIa non-small cell lung cancer: Long-term results. Am J Clin Oncol 1994;17:26–36.

25 Rosell R: A randomized trial comparing preoperative chemotherapy plus surgery with surgery alone in patients with non-small cell lung cancer. N Engl J Med 1994;330:153.

26 Roth JA, Fosella F, Komaki R, Ryan MB, Putnam JB, Lee JS, Dhingra H, De Caro L, Chasen M, MacGavran M, Atkinson EN, Hong WK: A randomized trial comparing perioperative chemotherapy and surgery with surgery alone in resectable stage IIIa non-small cell lung cancer. J Natl Cancer Inst 1994;86:673.

27 Pass HI, Pogrebniak HW, Steinberg SM, Mulshine J, Miuna J: Randomized trial of neoadjuvant therapy for lung cancer: Interim analysis. Ann Thorac Surg 1992;53:992–998.

28 Weiden PL, Piantadosi S and The Lung Cancer Study Group: Preoperative chemotherapy (cisplatin and fluorouracil) and radiation therapy in stage III non-small cell lung cancer: A phase II study of the Lung Cancer Study Group. J Natl Cancer Inst 1991;83:266.

29 Weiden P, Piantadosi S: Preoperative chemotherapy (cisplatin and fluorouracil) and radiation therapy in stage III non-small cell lung cancer. Lung Cancer 1991;7(suppl):157.

30 Strauss GM, Herndon JE, Sherman DD, Mathisen DJ, Carey RW, Choi NC, Rege VB, Modeas C, Green MR: Neoadjuvant chemotherapy and radiotherapy followed by surgery in stage IIIa non-small cell carcinoma of the lung: Report of a Cancer and Leukemia Group B Phase II Study. J Clin Oncol 1992;10:1237.

31 Dillman RO, Seagren SL, Propert K, Guerra J, Eaton WL, Perry MC, Carey RW, Frei E 3rd, Green MR: A randomized trial of induction chemotherapy plus high-dose radiation versus radiation alone in stage III non-small cell lung cancer. N Engl J Med 1990;320:940–945.

32 Kreisman K, Lisbona A, Olson L, Propert K, Modeas C, Dillman R, Seagren S, Green MR: Effect of prestudy tumor stage and of chemotherapy-radiotherapy in non-small cell lung cancer. Chest 1990;98:107S.

33 Faber LP, Kittle CF, Warren WH, Bonomi PD, Taylor SG, Reddy S, Lee MS: Preoperative chemotherapy and irradiation for stage III non-small cell lung cancer. Ann Thorac Surg 1989;47:669.

34 Faber L, Bonomi O: Combined preoperative chemoradiation therapy. Chest Surg Clin North Am 1991;1:43–59.

35   Rusch VW, Albain KS, Crowley JJ, Rice TW, Lonchyna V, McKenna R, Lingston RB, Griffin BR, Benfield JR: Surgical resection of stage IIIa and stage IIIb non-small cell lung cancer after concurrent induction chemoradiotherapy: A Southwest Oncology Group Trial. J Thorac Cardiovasc Surg 1993; 105:97.

36   Albain KS, Rusch VW: Concurrent cisplatin/etoposide plus chest radiotherapy followed by surgery for stages IIIa (N2) and IIIb non-small cell lung cancer: Mature results of Southwest Oncology Group Phase II Study 8805. J Clin Oncol 1995;13:1880.

37   Mountain CF: Lung cancer staging: 1997 revisions. 2nd International Congress on Lung Cancer, Crete, Nov 9–13, 1996, pp 11–13.

38   Green M: New adjuvant strategies for the management of resectable non-small cell lung cancer. Chest 1993;103(suppl):352–355.

39   Trillet-Lenoir V, Arpin D, Mercatello A, Oskham R, Palmer PA, Souquet PJ, Champel F, Boyer J, Cordier JF, Franks CR: Combination IL-2 and cisplatinum: A promising treatment for bronchioloalveolar carcinoma? Eur J Cancer 1993;29A:1917–1918.

40   Vokes E, Hochster H, Lotze M, et al: Recombinant human interleukin-4 SCH 39400 in non-small cell lung cancer: Results of a phase II investigation. Lung Cancer 1994;11(suppl 1):129.

David W. Johnstone, MD, Division of Cardiothoracic Surgery,
University of Rochester Medical Center, Box Surgery,
601 Elmwood Avenue, Rochester, NY 14642 (USA)
Tel. (716) 275-1509, Fax (716) 273-1011, E-Mail djohnstone@ctsurg.urmc.rochester.edu

Schiller JH (ed): Updates in Advances in Lung Cancer. Prog Respir Res.
Basel, Karger, 1997, vol 29, pp 35–55

Chapter 3
·····················

Role of Chemotherapy and Radiation Therapy in the Treatment of Locally Advanced Non-Small Cell Lung Cancer

Minesh P. Mehta

Human Oncology, University of Wisconsin Medical School, Madison, Wisc., USA

Introduction

Nonmetastatic, locally advanced, unresectable or medically inoperable non-small cell lung cancer (NSCLC) afflicts approximately 60,000 individuals in the US on an annual basis [1]. Although in earlier stage disease resection produces the best survival, these patients, in general, are not candidates for surgical intervention. For several years, the mainstay of treatment in the US for these patients has been 60 Gy radiation therapy at 2 Gy/fraction over 30 treatment days. This therapeutic choice resulted from the Radiation Therapy Oncology Group (RTOG) protocol 7301 which demonstrated that as total dose was escalated from 40 to 60 Gy, tumor control and long-term survival (> 2 years) improved [2]. Unfortunately, this radiotherapeutic approach has failed to achieve cure in the substantial majority of patients, prompting a search for newer therapeutic alternatives.

The new therapeutic paradigms are based on analysis of failure patterns which suggest that the problem is at least bifactorial. Recent studies incorporating posttherapy bronchoscopic evaluation indicate that local failure after conventional radiotherapy remains problematic and may approach 80% [3]. Distant metastases are also identified in the majority of these patients at death. Both local and distant control therefore assume priority in the newer therapeutic strategies, which have therefore utilized radiation dose escalation and intensification with strategies such as hyperfractionation, acceleration, brachytherapy boost and three-dimensionsal (3-D)/conformal techniques, as well as systemic targeting with combination chemoradiotherapeutic approaches. In this chapter, the major clinical results in these areas are presented.

Preradiation Induction Chemotherapy

Biological Basis and Hypothetical Considerations for Chemoradiotherapy

Presurgical neoadjuvant chemotherapy is an active area of investigation and has been addressed in Chapter 2. Several large randomized trials have tested preradiation induction chemotherapy over the last 10 years. The hypothetical benefits of this approach include targeting potential sites of micrometastatic disease at the earliest clinically detectable phase, thereby accomplishing the goals of treating minimal disease and avoiding several generations of cell division and possible development of drug resistance. Also, there is the expectation that the primary site of disease may respond to cytotoxic agents, thereby improving the efficacy of subsequent radiation. Sequential delivery was adopted first to avoid overlapping toxicities and also to ensure that both modalities could be delivered near the planned maximal doses.

The biological basis for the interaction of chemotherapeutic agents and radiation has been summarized by Fu [4] who suggested that possible advantages would include: (1) true radiosensitization as seen by making the dose-response curve steeper; (2) decreasing sublethal and potentially lethal damage repair; (3) cell-cycle synchronization, resulting in accumulation of cells in more radiosensitive phases, and (4) reoxygenation of hypoxic tumor regions by decreasing bulk and improving vascular supply. Neither the optimal agent(s) nor the appropriate schedule have been fully defined. Only five agents not considered 'new' have objective response rates exceeding 15%. These include cisplatin, mitomycin C, ifosfamide, vindesine and vinblastine; carboplatin produces lower response rates, but prolongs survival [5]. Kallman [6] has investigated the sequencing question in a series of experiments using fractionated radiotherapy and various chemotherapy schedules in an in vivo murine tumor model. Using tumor growth delay as an endpoint, the maximal therapeutic effect was achieved by combining daily cisplatin with fractionated radiotherapy.

CALGB 8433: Randomized Trial of Radiotherapy Alone versus Chemoradiotherapy

The landmark clinical trial that has lifted induction chemotherapy from the investigative realm to clinically accepted standard practice was conducted by the Cancer and Leukemia Group B (CALGB) from 1984 to 1987, with 7-year follow-up demonstrating survival improvement [7]. Patients with stage III NSCLC (clinical T_3 or N_2, M_0, either IIIA or IIIB according to the current staging system) with favorable prognostic features including CALGB performance status 0 or 1, $< 5\%$ body weight loss and normal hematologic and serum chemistry parameters were randomized to 60 Gy alone or the same radiation therapy preceded by 2 cycles of vinblastine (5 mg/m^2) as a weekly

Table 1. CALGB 8433: 7-year survival analysis

	n	MS (Mo)	Percent survival by year of follow-up						
			1	2	3	4	5	6	7
Radiotherapy	77	9.6	40	13	10	7	6	6	6
Chemoradiotherapy	78	13.7	54	26	24	19	17	13	13

n = Number of patients; MS (Mo) = median survival in months.

intravenous bolus times 5 and cisplatin (100 mg/m^2) on days 1 and 29. One hundred and fifty-five eligible patients were evaluated, with overall and median survival for the combination arm demonstrating superiority over radiation therapy alone. For the entire cohort, median survival was 11 months; 13.7 months for the combination compared to 9.6 months for radiotherapy alone. The 7-year survival figures are presented in table 1 and indicate that after 5 years, the survival probability was 2.8 times greater for patients in the combination arm. Whereas the magnitude of benefit as a ratio is quite impressive, absolute numbers provide a very sobering picture. Only 14/78 patients from the combination arm and 5/77 in the radiation alone arm survived beyond 4 years. No survival benefit accrued from the combination arm for the 44 patients with large cell carcinoma.

RTOG 8808/ECOG 4588: Confirmatory Trial for CALGB 8433

This major intergroup randomized trial was conducted to not only verify the results observed in the CALGB 8433 trial, but also to explore the potential role of hyperfractionated radiation therapy. Between 1989 and 1992, 452 eligible patients, selected using criteria similar to the CALGB study, were randomized to standard radiotherapy (60 Gy in 30 fractions of 2 Gy each), hyperfractionated radiotherapy (69.6 Gy in 58 fractions of 1.2 Gy each, given twice daily), or standard radiation with the same preradiation chemotherapy as in CALGB 8433. In 1995, with a potential median follow-up of 33 months, the study results demonstrated the superiority of the combined modality arm over both radiation regimens [8]. The median and percent survival for the first 3 years are presented in table 2 and compared with CALGB 8433. The improvement in median survival in the combination arm to 13.8 months was not only comparable to the 13.7 months in the CALGB study, but was also superior to 11.4 months from 60 Gy and 12.3 months from 69.6 Gy. One-year survival evaluation confirmed the trend.

Table 2. RTOG 8808: 3-year survival analysis

	n	MS (Mo)	Percent survival by year		
			1	2	3
Radiotherapy 60 Gy (CALGB)	77	9.6	40	13	10
Radiotherapy 60 Gy (RTOG)	153	11.4	46	19	6
Radiotherapy 69.6 Gy (RTOG)	156	12.3	51	24	13
Chemoradiotherapy (CALGB)	78	13.7	54	26	24
Chemoradiotherapy (RTOG)	152	13.8	60	32	15

n = Number of patients; MS (Mo) = median survival in months.

Meta-Analysis of Randomized Combination Therapy Trials

Whereas the large randomized intergroup trial verified the initial survival benefit from combination therapy, it once again dramatically illustrated the minimal overall gains, as measured by the fact that by 3 years, more than 85% of all patients had died. Interestingly, it also raised the issue of altered fractionation, since the survival advantage from combination therapy appeared to have vanished by year 3 (3-year survival of 15% from the RTOG combination therapy arm compared to 13% from the RTOG hyperfractionation arm). This is of major concern, since both the CALGB and the intergroup trials were conducted in extremely well-selected patients from the large cohort of advanced IIIA and IIIB cases, and extrapolation of these results to patients with inferior prognostic features has never been validated in the context of randomized, controlled clinical trials, and yet appears to have become rather popular in the arena of day-to-day practice. Opponents of combination therapy regimens also point to a number of negative trials that have failed to demonstrate survival benefit from this approach. For example, in a recent large randomized trial of 302 patients from Sweden treated either with 56 Gy alone or 56 Gy preceded by 3 cycles of cisplatin (120 mg/m^2) and etoposide (100 mg/m^2 intra-venously for 3 days), no survival advantage was detected. This negative result is particularly alarming given the 80% increase in total cisplatin dose in comparison to the CALGB trial [9].

In this context, it is useful to review the three recently published meta-analyses that have attempted to address this question. In 1995, Marino et al. [10] analyzed 14 randomized trials comprising 1,887 patients, estimating survival from published reports, rather than re-analyzing all patient data points. This report was based on a Medline search in all languages, and all studies included IIIA/B patients and compared radiotherapy plus chemotherapy to radiotherapy alone. Ten of the 14 trials used a cisplatin-based chemotherapy regimen, and for this group, the estimated pooled odds ratio of death at 1 and 2 years was 0.76 and 0.70, compared to radiotherapy alone; this translates to a reduction in mortality at 1 and 2 years of 24 and 30%, respectively. For the noncisplatin regimens, the mortality reduction of 5 and 18% at 1 and 2 years was less impressive. Unfortunately, in keeping with the intergroup study findings, no significant survival advantage from combination therapy was detected at 3 and 5 years, implying that the modest survival benefit does not carry through past the second year of survival. These results are all the more alarming given that the intergroup study was not actually included in this meta-analysis.

In 1995, Stewart and Pignon [11] from the British Medical Research Council and the Institut Gustave Roussy, on behalf of the Non-Small Cell Lung Cancer Collaborative Group, published a meta-analysis report using updated data on 9,387 individual patients from 52 randomized clinical trials in order to evaluate the effect of cytotoxic chemotherapy on survival in patients with NSCLC. This analysis contained 22 trials with 3,033 patients comparing radical radiotherapy with combination regimens, 11 of which were cisplatin-based. The overall odds ratio of death of 0.9 indicated a 10% reduction in the risk of death for the combination regimens, with an overall absolute survival benefit at 2 and 5 years of a mere 3 and 2%. The strongest survival trend was observed in the cisplatin-based trials, with an odds ratio of death of 0.87 (13% reduction in risk of death) and an absolute survival benefit at 2 and 5 years of 4 and 2%.

The most recent of these meta-analyses was published by Buccheri and Ferrigno [12] in 1996, evaluating 17 randomized trials with 1,355 patients, all treated with cisplatin-based regimens. The mean radiotherapy dose was 54 Gy. Nine of the 17 trials demonstrated a survival benefit from the combination approach, whereas the other 8 did not. Overall survival was superior in the combination group with 1- and 2-year survival figures of 44 and 41% for the combination arms vs. 19 and 14% for the radiotherapy arms.

Other Chemoradiotherapy Combinations

Five major themes emerge from the induction chemotherapy trials: (1) a modest survival benefit accrues from this approach; (2) meta-analyses reveal that for the most part, this survival benefit is restricted to the first 2 years; (3) cisplatin-based regimens appear to be superior to other types of chemotherapy; (4) toxicities are not substantially enhanced, and (5) failure pattern analysis of these trials reveal mixed results; although there is some indication of a

Table 3. Concurrent chemoradiotherapy trials in advanced NSCLC

Year	Author	n	RT (Gy)	CT	RR (%)	MS (wk)	% 1 yr	% 2 yr	p
1988	Soresi	45	50.4	P qwk	64	69	–	–	NS
		50	50.4	–	50	47	–	–	
1992	Trovo	85	45	P qd	51	43	–	–	NS
		88	45	–	59	44	–	–	
1992	Schaake-Konig	102	55	P qd	90	–	54	26	0.04
		98	55	P qwk	75	–	44	19	NS
		108	55	–	70	–	46	13	NS
1995	Blanke	110	60–65	P	50	43	43	18	NS
		114	60–65	–	38	46	45	13	

n = Number of patients per arm; RT (Gy) = radiation therapy total dose; CT = type of chemotherapy; P qd = daily cisplatin; P qwk = weekly cisplatin; RR (%) = overall response rate; MS (wk) = median survival in weeks; % 1 and 2 yr = percent survival at 1 and 2 years; p = probability; NS = not statistically significant.

decrease in distant metastatic disease, locoregional failure remains a major issue, thereby reinforcing the notion that further approaches would have to control both local and systemic disease. Broadly speaking, these strategies fall into the categories of concomitant chemoradiotherapy or sequential chemoradio-therapy followed by chemotherapy dose intensification, i.e. 'consolidation'.

Concomitant Chemoradiotherapy
This approach was developed with the expectation that not only would distant micrometastases be eliminated, but that the problem of locoregional failure would also be addressed. Several cisplatin-based trials have been under-taken and are summarized in table 3 [13–16]. Three of the 4 trials failed to demonstrate a statistically significant survival advantage. The most widely quoted positive trial, reported by Schaake-Konig et al. [15], was conducted by the European Organization for Research and Treatment of Cancer (EORTC) and compared the addition of daily or weekly cisplatin to 55 Gy radiotherapy alone, with the daily cisplatin (6 mg/m^2) arm yielding a statistic-ally superior 2-year survival rate of 26% compared to 13% with 55 Gy alone. As expected, this approach resulted in net radiosensitization with improved locoregional control, but did not influence metastatic patterns. The time to local recurrence was significantly longer in the group receiving daily cisplatin. Daily cisplatin provided 'additional local control' at 1 and 2 years of 18 and

12%. Although this approach provides possible clues for future exploration, confirmatory evidence has not been forthcoming. A very similar trial by Trovo et al. [14], comparing 45 Gy with the same radiation plus daily cisplatin (6 mg/m^2), showed no differences in overall response, pattern of relapse or median survival, but doubled the incidence of grade 3 esophagitis.

The more aggressive concomitant regimens have attempted to combine conventional chemotherapy with radiation and several phase II reports have been published. In one such relatively mature study reported by the Southwest Oncology Group (SWOG 8805), the combination of concurrent cisplatin and etoposide with thoracic radiotherapy was associated with severe toxicity [17]. In a recent phase II study, 50 patients with inoperable stage III NSCLC were treated with concurrent chemoradiotherapy to determine the feasibility, toxicity, response rate, local control and survival of concurrent chemotherapy with cisplatin-etoposide and radiotherapy. Thoracic radiotherapy was administered to a total dose of 60 Gy. Concurrent chemotherapy consisted of cisplatin 20 mg/m^2/day plus etoposide 50 mg/m^2/day, from day 1 through day 5, every 4 weeks for 4 cycles. The overall response rate was an impressive 84%, including 68% complete response, providing proof of principle that such therapy can enhance local control. With a minimum follow-up of 23 months, overall survival was 70% at 1 year, 40% at 2 years and 35% at 3 years. Median survival was 18 months. Major hematologic toxicity occurred in 24% of the patients [18].

The major randomized trial using the more traditional concomitant approach reached a negative conclusion. In the Hoosier Oncology Group (HOG) trial, conventional radiation to 60–65 Gy was compared to the same radiation plus concurrent cisplatin at 70 mg/m^2 on days 1, 22 and 43. In a total cohort of 215 eligible patients, the median survival was 46 weeks on the radiation-alone arm and 43 weeks on the combined modality arm. The 1-, 2- and 5-year survival rates were 43, 18 and 5% on the combination arm versus 45, 13 and 2% on the XRT arm, respectively; no therapeutic advantage was identified [16].

Consolidation Chemoradiotherapy

The desire to utilize postradiation chemotherapy emerged from the recognition that a substantial majority of NSCLC patients fail at distant sites, in spite of up-front chemoradiotherapy. The strategy of 'consolidation' has therefore been incorporated in some of the combined modality trials, but has not been fully validated in a randomized context. Its value therefore remains questionable, particularly as significant toxicity questions remain.

Proof of principle was provided by a large randomized trial incorporating pre- and postradiation chemotherapy and reported by Le Chevalier et al. [3] in 1991. Three hundred and fifty-three patients were randomized to 65 Gy alone or the same radiotherapy preceded and followed by 3 cycles each of vindesine, lomustine, cisplatin and cyclophosphamide. One-, 2- and 3-year survival rates were 41, 14 and 4% for radiotherapy alone, compared to 51, 21 and 12% for the combined modality arm. The relative distant metastases rate in the radiotherapy-only arm was twice as high as that observed in the combination arm, thus validating the hypothesis. Unfortunately, local control at 17 and 15% was poor in both arms, and the survival figures were not impressively superior to other combination trials not using consolidation.

In a phase II feasibility study of consolidative chemotherapy, the CALGB indicated that dose intensity was severely compromised by compliance and toxicity issues [19]. In a 1996 report, Ajlouni et al. [20] tested the consolidation concept in a phase II trial utilizing chemotherapy concomitantly with accelerated radiotherapy followed by 2 cycles of high-dose cisplatin (120 mg/m^2) and vinblastine. They achieved an impressive median and 3-year survival of 17.5 months and 20%. However, 74% of patients required chemotherapy dose reduction due to toxicities, and distant metastases developed in 60% of patients. Similarly, in a 121 patient North Central Cancer Treatment Group (NCCTG) phase III trial, when 60 Gy was randomly compared to the same radiation preceded by 2 cycles and followed for consolidative purposes by another 2 cycles of chemotherapy, no survival benefit was noted, but myelosuppression was more frequent in the combined modality arm [21]. Therefore, it is premature to conclude that major gains have been achieved from consolidation, and the issue still remains investigational.

Altered Fractionation

Introduction

Why is the benefit from chemotherapy so small in lung cancer? Using data from 353 randomized patients treated at the Institut Gustave Roussy in Paris, Arriagada and co-workers [3] evaluated patterns of relapse and concluded that, whereas a measurable reduction in distant metastases was achieved with combination therapies, local failure remained an extremely common and important consideration. Using posttreatment bronchoscopic evaluation, they were able to demonstrate that the actual complete response rate at the primary site was only 16.5%. Enhancing the control of local disease therefore must receive very high priority, and several clinical trials are currently undertaking this task, either by incorporating surgery into the regimen or by altering the radiotherapeutic strategy by escalating or by intensifying the radiation with strategies such as hyperfractionation, acceleration, brachytherapy boost, 3-D/conformal techniques and the use of radiosensitizers.

Biological Rationale

Of several possible radiotherapeutic approaches, altered fractionation has been evaluated most comprehensively. The biological rationale underscoring this approach is based on the recognition that the mechanisms for acute and chronic radiation toxicities are different, and in fact, acutely responding tissues behave in a fashion very similar to most rapidly proliferating tumors. Therefore, escalation in total dose is likely to enhance both tumor control as well as acute toxicities. Late toxicities appear to be significantly correlated to the fraction size, i.e., the amount of radiation per fraction. Therefore, dose intensification by increasing conventional daily fraction sizes from 1.8–2.0 Gy to, say for example, 3 Gy would enhance tumor control as well as both acute and chronic toxicities, thereby negating the improvement in the therapeutic index. Total dose can be escalated without increasing late toxicities by simply utilizing more daily single fractions; however, there is a rapid limit to this benefit as well as the risk that prolongation of the therapeutic duration may result in loss of tumor control. Hyperfractionation, utilizing multiple daily fractions, each typically smaller than the conventional 1.8-2.0 Gy fractions, spaced approximately 6–8 h apart and delivered over a total treatment duration of approximately 6 weeks, not dissimilar to conventional schedules, holds the promise of improving the therapeutic index by increasing the overall effective tumor dose without increasing late toxicities.

RTOG Hyperfractionation Trials

The RTOG has contributed substantially in designing and conducting prospective trials exploring hyperfractionation. Following their early studies in the 1970s which demonstrated substantial mucositis with a 1.5-Gy bid schedule, a dose-seeking phase I/II trial of 1.2 Gy bid was initiated. This large trial (RTOG 8311) evaluated total doses in the 60–79.2 Gy range in 884 patients. Of these, 350 met the criteria for CALGB 8433 and these patients had the best outcome in the 69.6-Gy arm of the study with a median 1- and 2-year survival of 13 months and 56 and 29% [22]. Parenthetically, comparable rates for the CALGB 8433 chemoradiotherapy arm were 13.7 months and 54 and 26%. When these two modalities were compared in a prospective randomized fashion in RTOG 8808 (table 2), preliminary results favored the combination arm, but this benefit did not persist beyond 2 years and outcomes from hyperfractionation and chemoradiotherapy appear very similar, prompting some very serious quality-of-life, toxicity and cost-effectiveness questions.

Accelerated Fractionation

A second major radiotherapeutic altered fractionation concept that has recently been extensively tested is based on the recognition from tumor cell kinetic studies that several primary lung carcinomas have extremely short (<5 to 7 days) potential doubling times [23]. The rapid repopulation from these tumor cells would have a major detrimental impact on conventional 6-week-long radiation schedules which would therefore start losing their effectiveness within a couple of weeks of initiation of therapy. In order to overcome this, it would be necessary to considerably shorten the overall treatment time, i.e. 'acceleration'. From a practical standpoint, the easiest way to accelerate is to use multiple daily fractions, <1.8–2.0 Gy each, but with a daily cumulative dose >1.8–2.0 Gy, a biologically comparable total dose, and a significantly shortened overall treatment duration. This concept therfore marries acceleration with hyperfractionation and is known as hyperfractionated accelerated radiotherapy (HART). When weekend breaks are eliminated so that there is no interruption, the word 'continuous' is added, resulting in the well-known acronym, CHART.

Clinical trials of CHART in NSCLC were first piloted in England by Saunders and Dische, and based on their encouraging preliminary data, a large, multicenter randomized European trial has just been completed [24]. A total dose of 54 Gy was delivered over 12 continuous treatment days at 1.5 Gy thrice daily, separated by intervals of 6 h, resulting in an 18-hour treatment day, weekends included. The initial results of this randomized trial demonstrate a statistically significant 10% survival benefit at 2 years in comparison to a standard 60-Gy schedule. The odds ratio, in favor of the CHART regimen, was 0.75, suggesting a 25% decline in the risk of death.

The CHART regimen has not gained popularity in the US because of manpower and other logistic constraints imposed by the lengthy treatment days. An alternative, more practical regimen spaced over 15 days, including 2 weekend breaks has been piloted by ECOG (ECOG 4593) with a median and 1-year survival of 13.5 months and 57% [25]. Although a direct comparison of these various trials is not statistically possible, a simple comparison is presented in table 4. Clearly, the data seem to suggest that the more aggressive treatment options, including combination therapy as well as altered fractionation, yield results superior to conventional, once-daily radiation.

Concomitant Boost

This technique of accelerating radiotherapy dose delivery reduces the overall treatment time by 1–2 weeks by giving the boost as a second fraction during the normal course of standard radiotherapy. Phase I and II studies have been carried out by the RTOG. In the largest such report, 355 patients received 1.8-Gy fractions to standard large fields, followed 4–6 h later by 1.8-Gy boost field radiation, given 2–3 times each week [26]. The total dose was escalated from 63 to 70.2 Gy in 5 weeks. Although there was some increase

Table 4. Survival comparison of selected radiotherapy, chemoradiotherapy and altered fractionation trials

Author	Study	Arm	Survival: median, % 1–3 years			
			M	1	2	3
Dillman	CALGB 8433	qd RT	9.6	40	13	10
Sause	RTOG 8808	qd RT	11.4	46	19	6
Schaake-Konig	EORTC	qd RT	–	46	13	2
Le Chevalier	IGR	qd RT	10	41	14	4
Saunders	CHART	qd RT	$\cong 12$	$\cong 50$	20	$\cong 10$
Cox	RTOG 8311	bid RT	13	56	29	$\cong 20$
Sause	RTOG 8808	bid RT	12.3	51	24	13
Saunders	CHART pilot	tid RT	$\cong 15$	64	34	$\cong 18$
Saunders	CHART	tid RT	$\cong 15$	$\cong 61$	30	$\cong 20$
Tannehill	ECOG 4593	tid RT	13.3	57	–	–
Dillman	CALGB 8433	CT + RT	13.7	54	26	24
Sause	RTOG 8808	CT + RT	13.8	60	32	15
Schaake-Konig	EORTC	RT + qwk CT	–	44	19	13
Schaake-Konig	EORTC	RT + qd CT	–	54	26	16
Le Chevalier	IGR	CT + RT	12	51	21	12

qd = Once daily; bid = twice daily; tid = thrice daily; qwk = once weekly; RT = radiotherapy; CT = chemotherapy; $\cong$ = estimated from survival curves; M = median survival in months.

in acute toxicity in the higher dose arm, late toxicities were not enhanced. Median survival remained at 9 months for the various cohorts, but in the high-dose arm, 1- and 2-year survival were 44 and 22%, comparable to what is contemporarily achieved with combination therapy. This approach has not been further tested in a randomized fashion.

Chemotherapy plus Altered Fractionation

Based on the principles and results outlined thus far, it is clearly logical to consider a combination of chemotherapy and altered fractionation. Several such phase I/II trials have now been reported and are summarized in table 5. The most impressive results in this setting were reported in an NCCTG study reported by Shaw et al. [27]. These investigators treated 23 patients with split-course hyperfractionated radiotherapy, using a 1.5 Gy bid to a toal of 30 Gy, followed by a 2-week break and an additional 30 Gy using a similar schedule. Two cycles of etoposide and cisplatin were given, one with commencement of

Table 5. Chemoradiotherapy trials with altered fractionation

Study	n	RT (Gy)	All bid	CT	Seq	Toxicity (%) bm	eso	lu	Survival other	M	1	2
Shaw	23	60 S	1.5	PE	C			26		26	74	51
R 9015	42	69.6	1.2	PV	C	45	24			12	54	28
R 9106	79	69.6	1.2	PE	C	57	53	25		19	67	35
Jeremic	61	64.8	1.2	–	–				G4 = 2	8	39	25
Jeremic	52	64.8	1.2	CaE	C				G4 = 4	18	73	35
Jeremic	56	64.8	1.2	CaE	C alt				G4 = 11	13	50	27
Pechoux	34	60	1.25	P Vd	C con				G5 = 2		53	33
Jeremic	66	69.6	1.2	–						14		
Jeremic	65	69.6	1.2	CaE	D					22		

n = Number of patients in each arm or study; RT (Gy) = total radiation dose in Gy; bid = twice daily (the numbers in this column represent fraction size in Gy); CT = chemotherapy; P = cisplatin; V = vinblastine; E = etoposide; Vd = vindesine; Ca = carboplatin; Seq = sequencing of chemotherapy and radiotherapy; C = concurrent; con = consolidation; D = daily with radiation; qwk = weekly; alt = every other week; Toxicity (%) = percentage of patients experiencing severe toxicity; bm = hematologic toxicity; eso = acute esophagitis; lu = acute pneumonitis; other = other toxicity; G5 = death; G4 = grade 4 toxicity; M = median survival in months; 1 and 2 = percent survival at 1 and 2 years.

the first radiation session, and the other when the second half of the radio-therapy course was commenced. Although toxicities were considerable with 26% grade 3 or greater acute pneumonitis, the overall median and 1- and 2-year survival figures were an impressive 26 months and 74 and 51%.

The RTOG has tested the 1.2-Gy bid regimen to 69.6 Gy with vinblastine (5 mg/m^2 weekly × 5) and cisplatin (75 mg/m^2 on days 1, 29 and 50) in their trial 90-15. Enhanced acute toxicities were substantial with 45% grade 4 or greater hematologic and 24% grade 3 or greater esophagitis. The median and 1- and 2-year survival was 12.2 months, 54 and 28%. In a subgroup with prognostic features similar to CALGB 8433, these survival figures were 17.5 months and 60 and 30% [28].

These rather excessive toxicities led to RTOG 91-06 which substituted etoposide in place of vinblastine, since this regimen had previously been tested and better tolerated in small cell lung cancer. Seventy-nine patients received 2 cycles of oral etoposide 100 mg/day, intravenous cisplatin 50 mg/m^2 on days 1 and 8 and hyperfractionated radiation therapy to 69.6 Gy. The median survival for patients comparable to CALGB 8433 was an impressive 21 months with 1- and 2-year survival figures of 70 and 42% (these values for CALGB 8433 were 13.7 months and 54 and 26%). Unfortunately, the associated toxicities from this regimen were also substantial with 57% grade 4 hematologic toxicity, 53% ≥ grade 3 esophagitis and 25% ≥ grade 3 pulmonary toxicity [29].

In a 34 patient phase II trial, the feasibility of combined concurrent hyperfractionated radiotherapy (60 Gy in 48 fractions of 1.25 Gy, twice daily) and chemotherapy consisting of cisplatin (6 mg/m^2 every day of radiotherapy) and vindesine (2.5 mg/m^2 once weekly) was tested. After a 3-week rest period, 2 full cycles of cisplatin (120 mg/m^2 at weeks 10 and 14) and vindesine (2.5 mg/m^2 at weeks 11, 12 and 13) were given. Treatment evaluation with thoracic computed scan, bronchoscopy, and bronchial biopsies was performed 3 months after completion of radiation therapy. Failure rates were estimated using a competing risk approach. The complete response rate was 50%. Local failure rates at 1 and 3 years were 53 and 56%, respectively. Distant metastases rates at 1 and 3 years were 27 and 29%. Overall survival rates at 1, 2 and 3 years were 53, 33 and 12%, respectively. Severe esophagitis was observed in 3 patients (9%). Lethal toxicity was observed in 2 patients. This phase II trial confirmed the feasibility of this type of approach and suggested that it may improve local control compared to conventional approaches [30].

The enhanced toxicity question from combining chemotherapy with altered fractionation has been investigated in a multicenter randomized trial using accelerated radiotherapy with or without concurrent carboplatin. One hundred patients with NSCLC were randomized to receive 1 of 4 treatments: arm 1, radiotherapy 60 Gy in fractions in 6 weeks; arm 2, accelerated radiotherapy 60 Gy in 30 fractions in 3 weeks; arm 3, radiotherapy as in arm 1 plus carboplatin 350 mg/m^2 during weeks 1 and 5 of radiotherapy; arm 4, radiotherapy as in arm 2 plus carboplatin 350 mg/m^2 during week 1. The median survival for all patients was 17.1 months with 33% survival at 2 years. The major toxicities were hematologic and esophageal. Patients receiving carboplatin had more neutropenia (p < 0.0001) and thrombocytopenia (p = 0.002) than patients receiving radiotherapy alone, and this was most marked in patients on arm 3. Both carboplatin and accelerated radiotherapy separately caused more severe esophagitis when compared to conventional radiotherapy alone (p = 0.011 and p = 0.0017, respectively). Esophagitis was more prolonged in patients having accelerated radiotherapy (p < 0.0001, median duration 3.2 months compared with 1.4 months for patients receiving conventional fractionation). Six patients (23%) treated on arm 2 required dilatation of esophageal stricture, 1 dying with a laryngoesophageal fistula. Clearly, such treatment schemes are associated with increased toxicities [31].

The first statistically significant survival advantage (median survival 34 vs. 77 weeks; p = 0.003) from chemo-hyperfractionated radiotherapy was reported by Jeremic and Shibamoto [32] in a three-arm randomized trial in the group receiving 100 mg/day carboplatin on days 1 and 2 with 100 mg etoposide on days 1–3 of each week during the course of radiotherapy (64.8 Gy; 1.2 Gy twice daily) in comparison to the same radiation alone. However, both acute and late toxicities were increased with this approach. Grade 4 acute toxicities were seen in 2, 4 and 11% of patients receiving radiation alone, radiation with weekly chemotherapy and radiation with chemotherapy on alternate weeks; the late toxicity values were 2, 4 and 9%.

Jeremic et al. [33] then conducted a subsequent phase III follow-up study with an interesting design change. To investigate the efficacy of concurrent hyperfractionated radiation therapy and low-dose daily chemotherapy in stage III NSCLC, 131 patients were randomly treated as follows: group I, 1.2 Gy twice daily to 69.6 Gy, and group II, same radiation with 50 mg of carboplatin and 50 mg of etoposide given on each day of radiotherapy. Group II patients had a significantly longer survival time than group I patients, with a median survival of 22 vs. 14 months and 4-year survival rates of 23 vs. 9% (p = 0.021). The median time to local recurrence and 4-year local recurrence-free survival rate were also significantly higher in group II than in group I (25 vs. 20 months and 42 vs. 19%, respectively, p = 0.015). In contrast, the distant metastasis-free survival rate did not significantly differ in the two groups. The two groups showed similar incidence of acute and late high-grade toxicity [33].

Whereas the results from Jeremic's trials are encouraging, further supportive evidence is necessary before combination chemo-altered fractionation radiotherapy can be recommended outside a protocol context. The RTOG has initiated an important three-arm randomized trial (RTOG 94-10) comparing the 'gold standard' of sequential chemoradiotherapy with concurrent chemoradiotherapy in one experimental arm and concurrent chemoradiotherapy with twice daily radiation in the other arm, and results of this trial are eagerly awaited.

New Horizons

3-D Conformal Radiation Therapy

Based on the well-documented knowledge that local relapse is the predominant mode of failure in lung cancer, it is logical to attempt to increase total dose to overcome this problem. The advent of sophisticated computers for treatment planning, advanced software and the widespread reliance on planning CT scans has spawned this entirely new field of 3-D conformal radiation therapy. The principles are elegantly simple: precise anatomic delineation of the target and identification of surrounding normal critical structures permits the design of multiple radiation therapy portals or fields approaching the tumor from almost any theoretical direction, thereby ensuring non-coplanarity. Theoretical modeling studies comparing such 3-D plans with conventional (or 2-D) plans have indicated the potential for escalating dose. This strategy is currently in its infancy, and only a handful of preliminary clinical investigations have been reported.

There are few data that unequivocally demonstrate clear survival advantage with radiation dose escalation. In part, this is a reflection of the technological inability in the past to deliver relatively high doses to the thorax. This dose-survival hypothesis therefore can only be partially answered with current data. To determine if there is such an effect, the survival of 941 patients in each of three schedule groups (20 Gy in 5 fractions; 30 or 36 Gy in 10 or 12 fractions, and 60 Gy in 30 fractions) was analyzed, before and after adjusting for the major prognostic factors, performance status and weight loss. The survival of patients planned to receive 60 Gy was significantly better than for patients planned to receive lower doses ($p < 0.0001$) with median survival increasing from 6.1 to 9.2 and 14.5 months for the 20-Gy, 30- or 36-Gy and 60-Gy groups, respectively. After adjusting for the effect of performance status and weight loss, death rates relative to the 20-Gy group were 79% for patients planned to receive 30 or 36 Gy and 53% for patients planned to receive 60 Gy. These data support the hypothesis that the increased survival in patients with NSCLC treated with higher dose radiotherapy is not due purely to patient selection [34].

Graham et al. [35] have reported on 70 patients treated prospectively with 3-D conformal techniques to doses in the range of 60–74 Gy; for early stage disease (medically inoperable stage I and II), the 2-year cause-specific survival rate is an astounding 90%, providing proof of principle that dose escalation may indeed be of value. For stage III patients, the 2-year cause-specific survival was 53%.

In a series of 45 patients treated with similar techniques at the Memorial Sloan-Kettering Cancer Center, the median and 2-year survival were 16.5 months and 33% [36]. Intrathoracic progression was observed in only 27% of patients, in contrast to the almost 80% expected from standard radiation. In another series of 37 patients with stage III NSCLC treated with high-dose (range 60–70 Gy) radiotherapy using conformal techniques, the median, 1- and 2-year survival rates for the entire group were 19.5 months, 75 and 37%, respectively; results which compare favorably with trials of chemoradiotherapy. However, local progression remained a significant problem in this trial as well, despite median radiotherapy doses of 66 Gy, indicating the necessity to explore higher doses [37]. Hazuka et al. [38] have reported the largest experience with this technique from the University of Michigan. In 88 patients, they have been able to escalate the dose to 74 Gy, with median and 2-year survival in the entire cohort being 15 months and 37%. Interestingly, as far as local progression-free survival and overall survival is concerned, a dose-response relationship was observed above 67.6 Gy within the stage III subgroup. Despite these doses, disease progression within the irradiated field was the predominant first site of treatment failure. Based on this preliminary work, they have initiated a multi-institutional dose-escalation trial and doses in excess of 80 Gy have already been safely delivered. Spurred by this exciting prospect of substantial increase in dose, the RTOG has initiated a multi-institutional phase I/II trial with doses in the range of 64.5–90.3 Gy.

Endobronchial Boost

This technique has been utilized for several years to boost localized endo-bronchial and peribronchial disease by endoscopically placing catheters in the tracheobronchial tree which can subsequently be loaded with radioactive isotopes. It has a well-defined role in the palliation of malignant airway occlu-sion, but its up-front incorporation as a dose-escalation technique awaits completion of better designed clinical trials.

The potential for this therapy was illustrated in a recent 41 patient pro-spective phase II study done to investigate the results of combined 40 Gy external beam and 25 Gy intraluminal radiotherapy in roentgenographically occult inoperable endobronchial carcinoma. In 39 evaluable patients, no bron-choscopically identifiable tumor was present at completion of therapy, for an overall bronchoscopic complete response rate of 39/41 (95%). With a median follow-up of 24.5 months, recurrence occurred in 2 cases [39].

In more advanced disease, preliminary experience is being gathered. In one such study, 62 patients with medically inoperable or surgically unresectable NSCLC were treated with both external beam radiotherapy and high-dose rate endobronchial brachytherapy. Treatment consisted of external beam radio-therapy (5,000–6,000 cGy in 5–6$\frac{1}{2}$ weeks) and weekly high-dose rate brachytherapy (3–5 fractions, 500 cGy at 1 cm). Overall median survival was 13 months. As expected, absence of nodal disease, which parenthetically cannot be targeted by endobronchial radiation, predicted for more favorable outcome with a median survival of 17 months. Clearly, additional studies are warranted to further investigate the use of this modality in the treatment of lung cancer [40].

Combinations of New Agents and Radiation Therapy

There is clearly a need to explore alternative therapeutic strategies in this disease, where success has been achieved only in small measure. Chapters 4–7 in this volume explore many of these newer strategies. As new drugs and drug combinations are being explored, the potential integration with radiation becomes a logical question. Several of these agents hold the promise of acting as radiosensitizing agents and preliminary phase I studies are currently underway. Two of the agents holding out the most promise in this category are paclitaxel and topotecan.

Paclitaxel is a microtubule-stabilizing agent that selectively blocks cells in G2 and M which are the most radiosensitive phases of the cell cycle. On the rationale that it could function as a cell-cycle selective radiosensitizer, Tischler et al. [41] examined the effects of combined drug-radiation exposures on human cancer cell lines, demonstrating dramatic interaction between paclitaxel and radiation. The sensitizer enhancement ratio (SER) for 10 nM paclitaxel at 10% survival was 1.8. Both paclitaxel and its analogue docetaxel are currently being actively tested in phase I and II trials. Activity has clearly been demonstrated in NSCLC, but appropriate combinations with radiation therapy have not yet been thoroughly investigated.

Another interesting class of compounds with potential radiosensitizing properties is the Topoisomerase I inhibitors, analogues of camptothecin such as topotecan, irinotecan (CPT 11) and 9-aminocamptothecin (9-AC). These drugs inhibit Topoisomerase I which cleaves single-stranded DNA. The potential radiosensitizing mechanisms include conversion of radiation-induced single-strand breaks into double-strand breaks and making these strand breaks irreversible. Preliminary phase I combinations of conventional radiation and concurrent topotecan are underway, but results are too preliminary to comment upon at present [42].

Interest in hypoxic cell sensitizers, in conjunction with altered fractionation, has been rekindled by recent laboratory and clinical experiments combining continuous hyperfractionated accelerated radiotherapy with nicotinamide, a vasodilator which modulates acute hypoxia and carbogen which modulates chronic hypoxia. A meta-analysis of hypoxic sensitizer studies by Overgaard and Horsman [43] included 10,703 cases entered into 83 randomized controlled trials and showed an overall improvement in local tumor control of 4.6% (p=0.00001) and in survival of 2.8% (p=0.005). Thus, ARCON (accelerated radiotherapy, carbogen and nicotinamide) has been introduced in several European clinical trials. The use of ARCON recognizes that tumor cell proliferation is an important cause of failure, in addition to hypoxia. Preliminary application in lung cancer is currently being studied [44].

Conclusions

The most recent US projections indicate that, in keeping with the decline in tobacco consumption, the overall incidence of lung cancer has started to decline. However, a substantial proportion of these patients still present with locoregionally advanced disease. Although the overall prognosis for most of these patients remains dismal, concerted effort over the last two decades has yielded some very interesting and promising advances, based primarily on an understanding of the disease process.

Early therapeutic strategies were predicated on the assumption that local control was of paramount significance, high rates of local control could be achieved with conventionally fractionated radiation to 60 Gy, and that chemotherapy was minimally active, with no impact on micrometastatic disease. Randomized induction chemoradiotherapy trials belied this assumption by demonstrating a small but measurable survival gain, resulting primarily from control of metastatic disease. Simultaneously, better assessment of local control indicated that radiographic evaluation was extremely inaccurate and falsely inflated previously reported control rates, which in reality hovered around 20%. This observation led to renewed efforts at controlling the disease locally, and of several such strategies, altered fractionation has been demonstrated in controlled randomized trials to provide statistically significant and meaningful survival benefit. Combinations of chemotherapy and altered fractionation therefore represent the 'new frontier', and although it is too preliminary to judge its overall impact, this strategy has certainly resulted both in increased toxicity as well as improved survival in preliminary phase III trials.

Are we reaching a therapeutic ceiling? Future progress in this disease must balance toxicities with outcome, measured not just in survival terms, but also as it pertains to quality-of-life. As health care reform policies continue their increasing grip on medical decisions, the social impact in terms of cost-effectiveness will become a critical question. For example, how justifiable is it to treat the majority of advanced, nonmetastatic lung cancer patients with poor prognostic features with aggressive therapies when they have been excluded from the randomized trials? If the hyperfractionation results hold up as being equivalent to combination therapy results, should one or the other strategy be abandoned? Will new advances in terms of drug development, gene therapy, etc., make these issues redundant?

References

1   Bonomi P: Combined modality treatment for stage III non-small cell lung cancer. Lung Cancer 1995;12(suppl 2):41–52.
2   Perez CA, Pajak TF, Rubin P, Simpson JR, Mohiuddin M, Brady LW, Perez-Tamayo R, Rotman M: Long-term observations of the patterns-of-failure in patients with unresectable non-oat cell carcinoma of the lung treated with definitive radiotherapy. Report by the Radiation Therapy Oncology Group. Cancer 1987;59:1874–1881.

3   Le Chevalier T, Arriagada R, Quoix E, Ruffie P, Martin M, Tarayre M, Lacombe-Terrier MJ, Douillard JY, Laplanche A: Radiotherapy alone vs. combined chemotherapy and radiotherapy in nonresectable non-small cell lung cancer: First analysis of a randomized trial in 353 patients. J Natl Cancer Inst 1991;83:417–423.

4   Fu KK: Biological basis for the interaction of chemotherapeutic agents and radiation therapy. Cancer 1985;55:2123–2130.

5   Kris M, Cohen E, Gralla R: An analysis of 134 phase II trials in non-small cell lung cancer. Proceedings of the World Congress on Lung Cancer, Toronto, 1989, p 39, abstract 233.

6   Kallman RF: The importance of schedule and drug dose intensity in combinations of modalities. Int J Radiat Oncol Biol Phys 1994;28:761–771.

7   Dillman RO, Herndon J, Seagren SL, Eaton WL Jr, Green MR: Improved survival in stage III non-small cell lung cancer: Seven-year follow-up of cancer and leukemia group B (CALGB) 8433 trial. J Natl Cancer Inst 1996;88:1210–1215.

8   Sause WT, Scott C, Taylor S, Johnson D, Livingston R, Komaki R, Emami B, Curran WJ, Byhardt RW, Turrisi AT, Dar R, Cox JD: Radiation Therapy Oncology Group (RTOG) 88-08 and Eastern Cooperative Oncology Group (ECOG) 4588: Preliminary results of a phase III trial in regionally advanced, unresectable non-small cell lung cancer. J Natl Cancer Inst 1995;87: 198–205.

9   Brodin O, Nou E, Mercke C, Linden CJ, Lundstrom R, Arwidi A, Brink J, Ringborg U: Comparison of induction chemotherapy before radiotherapy with radiotherapy only in patients with locally advanced squamous cell carcinoma of the lung. Eur J Cancer 1996;32:1893–1900.

10   Marino P, Preatoni A, Cantoni A: Randomized trials of radiotherapy alone vs. combined chemotherapy and radiotherapy in stages IIIa and IIIb non-small cell lung cancer. A meta-analysis. Cancer 1995;76:593–601.

11   Stewart JP, Pignon LA: Chemotherapy in non-small cell lung cancer: A meta-analysis using updated data on individual patients from 52 randomized clinical trials. Non-Small Cell Lung Cancer Collaborative Group. BMJ 1995;311:899–909.

12   Buccheri G, Ferrigno D: Therapeutic options for regionally advanced non-small cell lung cancer. Lung Cancer 1996;14:281–300.

13   Soresi E, Clerici M, Grilli R, Borghini U, Zucali R, Leoni M, Botturi M, Vergari C, Luporini G, Scoccia S: A randomized clinical trial comparing radiation therapy vs. radiation therapy plus cis-dechlorodiammine platinum (II) in the treatment of locally advanced non-small cell lung cancer. Semin Oncol 1988;15:20–25.

14   Trovo MG, Minatel E, Franchin G, Boccieri MG, Nascimben O, Bolzicco G, Pizzi G, Torretta A, Veronesi A, Gobitti C: Radiotherapy vs. radiotherapy enhanced by cisplatin in stage III non-small cell lung cancer. Int J Radiat Oncol Biol Phys 1992;24:11–15.

15   Schaake-Konig C, van den Bogaert W, Dalesio O, Festen J, Hoogenhout J, van Houtte P, Kirkpatrick A, Koolen M, Maat B, Nijs A, Renaud A, Rodrigus P, Schuster-Uitterhoeve L, Sculier JP, van Zandwijk N, Bartelink H: Effects of concomitant cisplatin and radiotherapy on inoperable non-small cell lung cancer. N Engl J Med 1992;326:524–530.

16   Blanke C, Ansari R, Mantravadi R, Gonin R, Tokars R, Fisher W, Pennington K, O'Connor T, Rynard S, Miller M: Phase III trial of thoracic irradiation with or without cisplatin for locally advanced unresectable non-small cell lung cancer: A Hoosier Oncology Group protocol. J Clin Oncol 1995;13:1425–1429.

17   Albain KS, Rusch VW, Crowley JJ, Rice TW, Turrisi AT III, Weick JK, Lonchyna VA, Presant CA, McKenna RJ, Gandara DR: Concurrent cisplatin-etoposide plus chest radiotherapy followed by surgery for stages IIIA (N2) and IIIB non-small cell lung cancer: Mature results of Southwest Oncology Group Phase II Study 8805. J Clin Oncol 1995;13:1880–1892.

18   Reboul F, Brewer Y, Vincent P, Chauvet B, Faure CF, Taulelle M: Concurrent cisplatin, etoposide and radiotherapy for unresectable stage III non-small cell lung cancer: A phase II study. Int J Radiat Oncol Biol Phys 1996;35:343–350.

19   Clamon G, Herndon J, Eaton W, Rosenman J, Maurer LH, Cooper MR, Green MR: A feasibility study of extended chemotherapy for locally advanced non-small cell lung cancer: A phase II trial of cancer and leukemia group B. Cancer Invest 1994;12:273–282.

20   Ajlouni M, Chapman R, Kim JH: Accelerated-interrupted radiation therapy given concurrently with chemotherapy for locally advanced non-small cell lung cancer. Cancer J Sci Am 1996;2:314–320.

21   Morton RF, Jett JR, McGinnis WL, Earle JD, Therneau TM, Krook JE, Elliott TE, Mailliard JA, Nelimark RA, Malsymiuk AW, Drummond RG, Laurie JA, Kugler JW, Anderson RT: Thoracic radiation therapy alone compared with combined chemoradiotherapy for locally unresectable non-small cell lung cancer. Ann Intern Med 1991;115:681–686.

22   Cox JD, Azarnia N, Byhardt RW, Shin KH, Emami B, Pajak TF: A randomized phase I/II trial of hyperfractionated radiation therapy with total doses of 60.0 Gy to 79.2 Gy: Possible survival benefit with greater than or equal to 69.6 Gy in favorable patients with Radiation Therapy Oncology Group Stage III non-small cell lung carcinoma: Report of Radiation Therapy Oncology Group 83-11. J Clin Oncol 1990;8:1543–1555.

23   Wilson GD, Sanders MI, Dische S, Daley FM, Robinson BM, Martindale CA, Joiner B, Richman PI: Direct comparison of bromodeoxyuridine and Ki-67 labeling indices in human tumors. Cell Prolif 1996;29:141–152.

24   Saunders MI, Dische S, Barrett A, Parmar MKB, Harvey A, Gibson D: Randomized multicentre trials of chart vs. conventional radiotherapy in head and neck and non-small cell lung cancer: An interim report. Br J Cancer 1996;73:1455–1462.

25   Tannehill SP, Froseth C, Wagner H, Petereit DP, Mehta MP: A multi-institutional phase II study of hyperfractionated accelerated radiation therapy for unresectable non-small cell lung cancer. Initial report of ECOG 4593. Int J Radiat Oncol Biol Phys 1996;361(suppl):207.

26   Byhardt RW, Pajak TF, Emami B, Herskovic A, Doggett RS, Olsen LA: A phase I/II study to evaluate accelerated fractionation via concomitant boost for squamous, adeno and large cell carcinoma of the lung: Report of Radiation Therapy Oncology Group 84-07. Int J Radiat Oncol Biol Phys 1993; 26:459–468.

27   Shaw EG, McGinnis WL, Jett JR, et al: Pilot study of accelerated hyperfractionated thoracic radiation therapy plus concomitant etoposide and cisplatin chemotherapy in patients with unresectable stage III non-small cell carcinoma of the lung. J Natl Cancer Inst 1993;85:321–323.

28   Byhardt RW, Scott CB, Ettinger DS, Curran WJ, Doggett RL, Coughlin C, Scarantino C, Rotman M, Emami B: Concurrent hyperfractionated irradiation and chemotherapy for unresectable non-small cell lung cancer: Results of Radiation Therapy Oncology Group (RTOG) 90-15. Cancer 1995; 75:2337–2344.

29   Lee JS, Scott C, Komaki R, Fossella FV, Dundas GS, McDonald S, Byhardt RW, Curran WJ Jr: Concurrent chemoradiation therapy with oral etoposide and cisplatin for locally advanced inoperable non-small cell lung cancer: Radiation Therapy Oncology Group Protocol 91-06. J Clin Oncol 1996; 14:1055–1064.

30   Le Pechoux C, Arriagada R, Le Chevalier T, Bretel JJ, Cosset BP, Ruffie P, Baldeyrou P, Grunewald D: Concurrent cisplatin-vindesine and hyperfractionated thoracic radiotherapy in locally advanced non-small cell lung cancer. Int J Radiat Oncol Biol Phys 1996;35:519–525.

31   Ball D, Bishop J, Smith J, Crennan E, O'Brien P, Davis S, Ryan G, Joseph D, Walker Q: A phase III study of accelerated radiotherapy with and without carboplatin in non-small cell lung cancer: An interim toxicity analysis of the first 100 patients. Int J Radiat Oncol Biol Phys 1995;31:267–272.

32   Jeremic B, Shibamoto Y: Pre-treatment prognostic factors in patients with stage III non-small cell lung cancer treated with hyperfractionated radiation therapy with or without concurrent chemotherapy. Lung Cancer 1995;13:21–30.

33   Jeremic B, Shibamoto Y, Acimovic L, Milisavljevic S: Hyperfractionated radiation therapy with or without concurrent low dose daily carboplatin/etoposide for stage III non-small cell lung cancer: A randomized study. J Clin Oncol 1996;14:1065–1070.

34   Ball D, Matthews J, Worotniuk V, Crennan E: Longer survival with higher doses of thoracic radiotherapy in patients with limited non-small cell lung cancer. Int J Radiat Oncol Biol Phys 1993; 25:599–604.

35   Graham MV, Purdy JA, Emami B, Matthews JW, Harms WB: Preliminary results of a prospective trial using three-dimensional radiotherapy for lung cancer. Int J Radiat Oncol Biol Phys 1995;33: 993–1000.

36   Armstrong J, Zelefsky M, Burt M, Leibel S, Fuiks Z: Promising results for 3-dimensional conformal radiation therapy for NSCLC. Proc Am Soc Clin Oncol 1994;13:335.
37   Sibley GS, Mundt AJ, Shapiro C, Jacobs R, Chen G, Weichselbaum R, Vijayakumar S: The treatment of stage III non-small cell lung cancer using high-dose conformal radiotherapy. Int J Radiat Oncol Biol Phys 1995;33:1001–1007.
38   Hazuka MB, Turrisi AT III, Lutz ST, Martel MK, Ten Haken RK, Strawderman M, Borema PL, Lichter AS: Results of high-dose thoracic irradiation incorporating beam's eye view display in non-small cell lung cancer: A retrospective multivariate analysis. Int J Radiat Oncol Biol Phys 1993;27: 273–284.
39   Saito M, Yokoyama A, Kurita Y, Uematsu T, Miyao H, Fujimori K: Treatment of roentgenographically occult endobronchial carcinoma with external beam radiotherapy and intraluminal low dose rate brachytherapy. Int J Radiat Oncol Biol Phys 1996;34:1029–1035.
40   Aygun C, Weiner S, Scariato A, Spearman D, Stark L: Treatment of non-small cell lung cancer with external beam radiotherapy and high-dose rate brachytherapy. Int J Radiat Oncol Biol Phys 1992;23:127–132.
41   Tischler RB, Schiff PB, Geard CR, Hall EJ: Taxol: A novel radiation sensitizer. Int J Radiat Oncol Biol Phys 1992;22:613–617.
42   Berlin JD, Schiller JH, Hutson PR, Feierabend C, Simon K, Alberti D, Arzoomanian RZ, Boothman DA, Planchon S, Wuerzberger S, Glen V, Mehta MP, Wilding G: Phase I clinical and pharmacokinetic study of daily topotecan with thoracic irradiation. Proc Am Assoc Cancer Res 1996;37:164.
43   Overgaard J, Horsman MR: Modification of hypoxia-induced radioresistance in tumors by the use of oxygen and sensitizers. Seminars in Rad Oncol 1996;6:10–21.
44   Saunders M, Dische S: Clinical results of hypoxic cell radiosensitisation from hyperbaric oxygen to accelerated radiotherapy, carbogen and nicotinamide. Br J Cancer 1996;74(suppl 27):271–278.

Minesh P. Mehta, MD, Associate Professor of Human Oncology, University of Wisconsin Medical School, 600 Highland Avenue K4/B, Madison, WI 53792-0600 (USA)
Tel. (608) 263-8500, Fax (608) 263-9167, E-Mail mehta@humonc.wisc.edu

Schiller JH (ed): Updates in Advances in Lung Cancer. Prog Respir Res.
Basel, Karger, 1997, vol 29, pp 56–72

Chapter 4
....
Treatment of Stage IV Non-Small Cell Lung Cancer

Philip D. Bonomi

Rush-Presbyterian-St. Luke's Medical Center, Section of Oncology,
Chicago, Ill., USA

Introduction

Lung cancer continues to be the leading cause of the cancer-related deaths in both men and women in the United States. Approximately 85% of all lung tumors are classified as non-small cell carcinomas, and almost 90% of non-small cell lung cancer (NSCLC) patients will either present with metastatic disease or well-developed disseminated disease subsequently. Theoretically all of these patients are candidates for systemic therapy. During the period from 1975 to the present, investigators have conducted trials which have addressed a variety of questions in this group of patients. The results and ramifications of these studies will be summarized in this review.

Early Trials: What Is the Best Regimen?

During the 1970s, a number of investigators had observed relatively high remission rates with combination chemotherapy regimens in testicular cancer, Hodgkin's disease, non-Hodgkin's lymphoma, and breast cancer. When designing a combination regimen, chemotherapists attempted to apply the following principles: (1) selection of active single agents; (2) selection of agents whose major toxicities were non-overlapping, and (3) selection of agents with different mechanisms of antineoplastic activity [1]. Although lung cancer investigators were somewhat limited by the fact that the tumor was relatively insensitive to chemotherapy, several combination chemotherapy regimens were developed and tested in the mid-1970s. One of these consisted of cyclophosphamide, doxirubicin, methotrexate, and procarbazine (CAMP regimen) [2] and the other consisted of methotrexate, doxirubicin, cyclophosphamide and lomustine (MACC regimen) [3]. Each regimen produced response rates ≥40% in relatively small, single-institution phase II trials. Subsequently, these regimens were tested in phase III [4, 5] trials which included significantly larger numbers of patients, and the response rate for each regimen was <20%.

In the mid-1970s, cisplatin was a relatively new agent which had produced relatively low response rates in NSCLC. However, Eagan et al [6] at Mayo Clinic tested a series of cisplatin-containing combination regimens and observed a response rate of approximately 40% with a regimen consisting of cyclophosphamide, doxirubicin and cisplatin. Like the MACC and CAMP regimen, this cisplatin-containing regimen was also tested in a phase III trial but it produced a response rate of approximately 25% [7].

Encouraged by initial reports of relatively high response rates for the CAMP regimen a number of investigators designed other cisplatin-containing regimens including mitomycin-vinblastine-cisplatin (MVP), vindesine-cisplatin, and etoposide-cisplatin. The Eastern Cooperative Oncology Group (ECOG) tested each of these regimens in phase III trials, but none emerged as clearly superior with respect to survival [5, 7]. The etoposide-cisplatin regimen produced the highest 1-year survival rate of all the cisplatin- and noncisplatin-containing regimens which were tested by ECOG, but the differences in the 1-year survival rate for etoposide-cisplatin compared to the other regimens were not statistically significant [8]. The mitomycin-vinblastine-cisplatin regimen produced the highest response rate in three consecutive trials [5, 7, 9], and in two of these trials [7, 9] the response rate for MVP was statistically significantly higher than the other regimens [5, 7, 9]. Surprisingly, there was a trend for shorter survival in patients treated with the MVP regimen. The 1-year survival rate was 12% [8] and the median survival was consistently 4–6 weeks shorter than the other combination regimens (p=0.09) [5, 7, 9].

These observations emphasize the need to conduct phase III trials in which the primary objective is to determine the effect of treatment on survival In summary, none of the combination regimens which were tested from the mid-1970s to the mid-1980s was associated with significantly longer survival [8]. The median survival was 6 months and the 1-year survival rate was 19% in 893 patients treated in two consecutive phase III trials in which nine regimens were evaluated [8]. The response rates for cisplatin containing regimens in ECOG trials were approximately 25%. Investigators within other cooperative groups observed similar results [10, 11].

Cisplatin Dose: How Much Is Enough?

In addition to performing trials to identify the most effective combination regimen, some investigators were studying the effect of cisplatin dose. In 1981, Gralla et al [12] reported results of a randomized trial in which vindesine was combined with two doses of cisplatin – 60 versus 120 mg/m^2. They observed no significant difference in response rate, but survival was significantly longer in responding patients treated with the higher dose of cisplatin. Although the overall survival for patients treated with low dose (60 mg/m^2) versus high dose (120 mg/m^2) was not reported, the results of this study served as the basis for using higher doses of cisplatin. Subsequently, two studies testing the question of cisplatin dose were done [13, 14]: in the first trial etoposide was combined with cisplatin given at the same doses used by Klastersky et al [13], and in the second trial, etoposide was combined with cisplatin given at the same doses used by Gralla et al [12], namely 60 versus 120 mg/m^2. No significant differences were observed in response rate or overall survival

More recently, the Southwestern Oncology Group (SWOG) completed a phase III trial in which the following regimens were compared: cisplatin 100 mg/m^2 every 28 days versus cisplatin 200 mg/m^2 every 28 days versus mitomycin combined with cisplatin 200 mg/m^2 every 28 days [14]. Response rates for cisplatin 100 versus 200 mg/m^2 were not significantly different. Similar to ECOG trials the mitomycin-containing regimen produced a significantly higher response rate. However, neither of the high-dose cisplatin regimens produced significantly longer survival Although some investigators continue to use relatively high doses of cisplatin in combination regimens, the results of randomized trials described above [13, 14] do not appear to support a dose of cisplatin > 60 mg/m^2.

Another Platinum Question: Carboplatin versus Cisplatin?

Carboplatin, a cisplatin analog, was introduced into clinical trials in the mid-1980s. Both the Acute Leukemia Group [15] and ECOG [9] conducted relatively large single-agent trials. In the Cancer and Leukemia Group B (CALGB) study, carboplatin produced a 16% response rate [15], while a 9% response rate was observed with carboplatin in the ECOG trial [9]. Surprisingly, carboplatin was associated with slightly longer survival than cisplatin-containing combination regimens in the ECOG trial

There is no question that the toxicity profiles for carboplatin and cisplatin are different with significantly more myelosuppression appearing with carboplatin compared to cisplatin and with neurotoxicity and renal toxicity being associated with cisplatin but not with carboplatin. Similar degrees of nausea and vomiting are observed with each agent.

The question of whether there is a difference in the efficacy has been addressed in a randomized trial conducted in Belgium [16]. In this study, etoposide given on the same dose and schedule was combined with carboplatin 300 mg/m^2 or cisplatin 100 mg/m^2. No significant differences in response rates or survival were observed in this trial suggesting that carboplatin and cisplatin have a similar activity in NSCLC. Surprisingly, there was significantly more hematologic toxicity with the cisplatin regimen. Perhaps this observation was due to the relatively high dose of cisplatin. It is assumed that the ratio of carboplatin to cisplatin dose is 4 to 1. If this assumption is correct, then for a cisplatin dose of 100 mg/m^2 the equivalent dose of carboplatin would have been 400 mg/m^2.

Is Combination Chemotherapy Better than Supportive Care Alone?

There have been a number of randomized trials [17–22] in which combination chemotherapy regimens were compared to supportive care only in patients who had advanced NSCLC. These trials have shown conflicting results. In some of the studies, treatment with chemotherapy was associated with significantly longer survival [17, 19, 20, 22], while in others no significant survival difference was observed [18, 21]. Recently, a meta-analysis has shown that the median survival was improved by 6 weeks and the 1-year survival rate is improved by approximately 10% in patients treated with chemotherapy [23]. Although the results of their analyses failed to identify a specific regimen which was superior [23], a large database study conducted by the SWOG suggested that patients treated with a cisplatin-containing regimen have significantly longer survival than patients treated with noncisplatin-containing regimens [24].

At this point it appears that treatment with platinum-containing combination chemotherapy is associated with modest improvement in survival in patients who have stage IV NSCLC. Two meta-analyses have disclosed a 6-week improvement in median survival [23, 25]. In one of these reports the survival advantage was limited to the first 6 months [25], while in the other the 1-year survival rate was also improved in patients treated with chemotherapy [23].

Is Treatment Worthwhile?

How do physicians view these results? A Canadian study in which medical oncologists, radiation oncologists, surgeons and pulmonologists were asked to complete a questionnaire regarding their opinions about chemotherapy in advanced NSCLC showed that recent publications showing a survival benefit from chemotherapy did not appear to influence their treatment recommendations [26]. The majority of these physicians did not treat their advanced NSCLC patients with chemotherapy despite the fact that they believed chemotherapy prolonged survival The authors concluded that personal beliefs rather than universal knowledge guided their treatment decisions.

Are Combination Regimens more Effective than Single Agents?

There have been at least ten relatively large trials in which an active single agent was compared to a combination regimen [9, 27–35]. Seven of these studies [9, 27–31, 33] were conducted with cytotoxic agents identified primary to 1990, and three more recent trials have included vinorelbine, a newer vinca alkaloid [32, 34, 35], which produced a response rate of approximately 30% in two relatively large phase II trials [36, 37]. In seven of the randomized studies the combination regimens were associated with significantly higher response rates [9, 28–30, 32, 33, 35]. However, the overall survival was not significantly improved by treatment with the combination regimens in six [9, 27–31] of the seven trials using older chemotherapeutic agents. The one trial which showed a significantly longer survival for the combination regimen involved teniposide combined with cisplatin versus teniposide alone [33]. Two similar trials in which a different topoisomerase II inhibitor, etoposide, was combined with cisplatin did not reveal superior survival for patients treated with the combination regimen [28, 29]. In one of these trials, etoposide-cisplatin was compared to etoposide [29], and in the other trial the two-drug combination was compared to cisplatin alone [29]. Based on the collective results of these trials [28, 29, 33], it seems unlikely that regimens consisting of topoisomerase II inhibitors plus cisplatin are significantly better than either single agent alone.

Surprisingly, while being associated with a significantly lower response rate, carboplatin alone was associated with a modest but statistically longer duration of survival compared to three cisplatin-based combination regimens [9]. Critics argue that this observation was due to a random occurrence rather than to the therapeutic effect of carboplatin and although the critics may be correct, it is interesting that the other single agent, iproplatin, which was included in this trial was associated with a median survival duration which was virtually identical to the survival observed in patients treated with the combination regimens [9]. In addition, although the MVP regimen produced a significantly higher response rate, it was associated with a trend for shorter

Table 1. Phase III trials of vinorelbine-cisplatin versus vinorelbine alone or versus cisplatin alone

Investigators	Regimen	Patients	Response, %	p	Survival, weeks	p
LeChevalier et al. [34]	Vinorelbine-cisplatin	206	30	<0.001	40	0.01
	Vinorelbine	206	12		31	
DePierre et al. [32]	Vinorelbine-cisplatin	121	43	<0.0001	33[1]	0.48
	Vinorelbine	119	16		32	
Wozniak et al. [35]	Vinorelbine-cisplatin	205	25	NA[2]	7[3]	
	Cisplatin	207	10		6	0.001

[1] Time to progression was significantly longer, 20 weeks for VP and 10 weeks for V (p=0.03).
[2] Not available.
[3] One-year survival rates, VP=33%, P=12%.

survival [9]. Both of the single agents, carboplatin and iproplatin, produced significantly less severe and life-threatening toxicity compared to the combination regimens.

Two possible explanations for the unexpected survival results are: first, cytotoxic treatments may exert a positive effect on survival by slowing tumor proliferation without inducing objective tumor regression, and second, toxicity from cytotoxic treatments may shorten survival without producing clearly definable lethal toxicity. Perhaps, most important, the unexpected survival results from this study emphasize the need for carefully conducted phase III trials.

More recently, vinorelbine has been evaluated in three trials comparing vinorelbine-cisplatin to either drug alone [32, 34, 35] (table 1). In two of the studies, vinorelbine-cisplatin was compared to vinorelbine alone [32, 34], and in the third trial vinorelbine-cisplatin was compared to cisplatin alone [35]. In each trial a significantly higher response rate was observed with the two drug combinations. In contrast, survival data have shown conflicting results in the trials in which the combination regimens were compared to vinorelbine alone. LeChevalier et al [34] observed significantly longer survival in patients who received vinorelbine-cisplatin versus those who received vinorelbine alone, while DePierre et al [32] noted no significant difference in overall survival [31]. However, DePierre et al [32] observed significantly longer time to progression in patients treated with vinorelbine-cisplatin compared to those treated with vinorelbine alone.

In the third study [35] both time to progression and overall survival were significantly longer for patients treated with vinorelbine-cisplatin compared to cisplatin alone. It should be noted that the trials [34, 35] which showed significantly longer survival for the combination regimens contained approximately 200 patients per treatment arm, while the study which showed no difference in overall survival contained 119 patients per treatment arm [32]. Based on the observation, it appears that the vinorelbine-cisplatin regimen is the first combination regimen which has been associated with significantly better survival than either single agent alone.

The Late 1980's: What to Do Next?

In the late 1980s it was becoming apparent that none of the widely tested combination regimens had produced a significant impact on survival The median survival for patients treated with cisplatin-containing combination regimens was approximately 6 months and the 1-year survival rate was 19% [8]. Faced with these disappointing results, many investigators decided to concentrate on testing new agents in previously untreated NSCLC patients. This work has resulted in the discovery of two new classes of active drugs, the taxanes and topoisomerase I inhibitors. In addition, potentially more active analogs of drugs which had shown activity in NSCLC were identified. Vinorelbine (described above) is an example of a promising new vinca compound, and gemcitabine which is an active new nucleoside analog, a class of drugs which previously had not shown activity in NSCLC.

Vinorelbine: A New Vinca Alkaloid

Vinca compounds are plant alkaloids which exert their cytotoxic effect by binding to and inhibiting the polymerization of tubulin which results in decreased formation of microtubules. Vinorelbine appears to have greater affinity for the mitotic spindle than the neural neurotubules, suggesting that it might produce less neurotoxicity than other vinca alkaloids. Two relatively large phase II trials of vinorelbine have been done – one in France [36] and the other in Japan [37]. The dose of vinorelbine in the French trial was 30 mg/m² given intravenously on a weekly basis and in the Japanese trial the dose was 25 mg/m²/week. Both groups of investigators observed similar response rates in previously untreated patients: French trial – 29% in 78 patients [36], Japanese trial – 37% in 87 patients [37].

Vinorelbine was also discussed in the previous section which described the results of studies in comparing combination regimens to single agents. In the larger phase III studies the response rate for vinorelbine was 12% [32, 34,

38]. Although the low response rates observed in the phase III trials are somewhat disappointing, it should be noted the median survival for vinorelbine alone is approximately 8 months [32, 34, 38] which compares favorably with the 6-month median survival that had been observed with combination regimens in earlier trials [8]. The relatively long survival duration for patients treated with vinorelbine alone suggests that this agent has a positive effect on survival However, the relatively long survival might be due to the effect of stage migration because approximately 40% of the patients treated in the phase III vinorelbine trials had stage IIIb disease [32, 34] while only stage IV patients were included in the earlier trials of combination chemotherapy regimens [8].

However, in one of the phase III studies, 143 patients were treated with vinorelbine alone and 68 patients were treated with 5-fluorouracil-leucovorin [38]. The response rate for vinorelbine was 12% while it was only 3% for 5-fluorouracil and leucovorin. In addition, survival was significantly longer in vinorelbine-treated patients; median survival for vinorelbine was 30 versus 22 weeks for 5-fluorouracil (p = 0.03). This observation represents the second example of a single agent's prolonging survival in advanced NSCLC patients. This observation combined with the report that vinorelbine-cisplatin produced superior survival compared to cisplatin alone [35] provides increasing evidence that vinorelbine has a positive effect on survival in advanced NSCLC patients.

Vinorelbine is also available in an oral formulation which was tested in 162 patients. The first 25 patients were treated with 100 mg/m^2/week. If no previous radiation therapy had been given, patients received 80 and 50 mg/m^2/week respectively. The response rate was 14% and the median survival was 29 weeks [39].

Taxanes

The taxanes represent a new class of drugs derived from several species of yew trees. Like vinca alkaloids, these agents also affect the mitotic spindle, but unlike vinca alkaloids, which cause disaggregation of the microtubules, the taxanes promote polymerization and inhibit depolymerization of tubulin.

Taxol was the first taxane to be tested in NSCLC. Investigators within the ECOG [40] and at M.D. Anderson [41] evaluated this agent in small phase II trials, and they observed a response rate of approximately 20%, and, perhaps more importantly, the 1-year survival rate was 40% (table 2). Both groups gave taxol every 3 weeks as a 24-hour infusion, but the doses were slightly different: ECOG – 250 mg/m^2 and M.D. Anderson – 200 mg/m^2. More recently, results of phase II trials using shorter infusion time for taxol have been reported. In an Australian study, 51 previously untreated patients received taxol 175 mg/m^2 as a 3-hour infusion every 3 weeks and the response rate was 10% [42]. In another trial a 36% response rate was observed in 36 chemotherapy-

Table 2. Phase II trials of taxol

Investigators	Dose, mg/m^2	Schedule	Patients	Response, %
Chang et al. [40]	250	24-hour infusion every 3 weeks	24	21
Murphy et al. [41]	200	24-hour infusion every 3 weeks	25	24
Millward et al. [42]	175	3-hour infusion every 3 weeks	51	10
Hainsworth et al. [43]	200	1-hour infusion every 3 weeks	36	31

Table 3. Phase II trials of taxotere

Investigators	Dose, mg/m^2	Schedule	Patients	Response, %
Francis et al. [44]	100	1-hour infusion every 3 weeks	29	38
Fossella et al. [45]	100	1-hour infusion every 3 weeks	39	33
Fossella et al. [46][1]	100	1-hour infusion every 3 weeks	42	21

[1] Only patients who had progressive disease after treatment with cisplatin-based chemotherapy were eligible for this study.

naive patients who were treated with taxol 200 mg/m^2 given over 1 h (table 2) [43]. The dose-limiting toxicities of taxol are myelosuppression, particularly granulocytopenia, and peripheral neuropathy. Early concerns about cardiac toxicity have not been confirmed in these studies. Similarly, the problem of severe allergic reactions reported in early trials appears to have been reduced significantly by pretreatment with corticosteroids and H$_1$ and H$_2$ blockers.

Taxotere is a semisynthetic taxane which is obtained from the needles of the European yew. This agent has also shown significant activity given at a dose 100 mg/m^2 intravenously over 1 h every 3 weeks. Results of phase II trials are summarized in table 3 [44, 45]. Taxotere, another taxane, produced response rates of 33 and 38% in phase II trials. In addition, it is interesting that a response rate of 21% was observed in 42 patients who had previously been treated with cisplatin-based chemotherapy. Response rates in previously treated non-small cell lung cancer patients have generally been quite low. In an ECOG trial, patients whose disease progressed on carboplatin or iproplatin were subsequently treated with MVP [9]. This regimen produced a 6% response in previously treated patients compared to a 20% response rate in untreated patients [9]. The 21% response rate with taxotere [45] suggests a lack of cross-resistance between taxotere and cisplatin making this two-drug combination an attractive regimen to test. This is currently being done in a phase III ECOG trial

Topoisomerase I Inhibitors

Topoisomerase I is an enzyme which produces single-strand DNA breaks allowing unwinding of the DNA ahead of the replication fork. The enzyme allows passage of the unbroken DNA strand past the break site, and it covalently binds to the broken strand and subsequently religates the cleaved DNA strand. It is known that topoisomerase inhibitors increase the quantity of convalently bound topoisomerase I to cleaved DNA strands, but it is believed that the primary cytotoxic effect is related to double-strand DNA breaks [47].

Irinotecan (CPT_{11}), a topoisomerase I inhibitor, was tested in a Japanese phase II trial in which the drug was given weekly at a dose of 100 mg/m². A 32% response rate was observed in 72 previously untreated NSCLC patients. Major toxicities consisted of leukopenia and diarrhea [48]. Cisplatin 80 mg/m² and escalating doses of irinotecan plus granulocyte colony-stimulating factor (G-CSF) have been combined in a phase I trial in NSCLC. The recommended dose for phase II testing is cisplatin 80 mg/m² and irinotecan 80 mg/m² on days 1, 8 and 15 plus G-CSF [49]. Diarrhea was the dose-limiting toxicity.

Gemcitabine

Gemcitabine, 2-deoxy-2,2-difluorocytidine, is an analog of cytosine arabinoside. Like its precursor, gemcitabine is converted to the triphosplate which when incorporated into DNA results in chain termination. Compared to cytosine arabinoside, larger amounts of gemcitabine accumulate intracellularly, which may be the basis for gemcitabine's broad spectrum of activity [47]. There have been four large phase II trials which evaluated gemcitabine in advanced NSCLC [50–53]. The results are depicted in table 4. Using doses of 800–1,250 mg/m² on days 1, 8 and 15 repeated every 28 days, four groups of investigators have observed response rates of 20%. The predominant toxicity with this agent is myelosuppression with thrombocytopenia being more prominent than leukopenia.

Table 4. Phase II trials of gemcitabine[1]

Investigators	Dose, mg/m²	Schedule	Patients	Response, %
Anderson et al. [50]	800–1,000	Days 1, 8, 15 every 4 weeks	79	20
Negoro et al. [51]	1,000	Days 1, 8, 15 every 4 weeks	189	22
Abratt et al. [52]	1,000–1,250	Days 1, 8, 15 every 4 weeks	76	20
LeChevalier et al [53]	1,250	Days 1, 8, 15 every 4 weeks	161	22

[1] Gemcitabine was given as a 30-min intravenous infusion.

New Drugs: What Is the Best Regimen?

Both vinorelbine and taxol have been combined with cisplatin and compared to other cisplatin-based regimens in relatively large phase III trials [34, 54, 55]. As discussed previously vinorelbine-cisplatin has been compared to vinorelbine alone in two French trials, and to cisplatin alone in a SWOG trial [35]. In one of the French trials, vinorelbine-cisplatin was also compared to a combination regimen which consisted of vindesine and cisplatin [34]. Vinorelbine-cisplatin produced a superior response rate and also was associated with significantly longer survival compared to vindesine-cisplatin [34]. This represents the first time that a cisplatin-containing combination regimen was associated with superior survival when compared to another cisplatin-containing regimen in the large phase III trial

Taxol combined with cisplatin has been compared to regimens which consisted of a topoisomerase II inhibitor plus cisplatin in two phase III trials [54, 55]. ECOG investigators compared taxol 135 mg/m² intravenously as a 24 hour infusion, plus cisplatin 75 mg/m² on day 2 versus taxol 250 mg/m² given as a 24-hour infusion combined with cisplatin 75 mg/m² on day 2, and G-CSF 5 mg/kg subcutaneously daily on days 3–10 to etoposide-cisplatin which was selected as the reference regimen because it had produced the highest 1-year survival rate in earlier ECOG trials. A total of 574 patients were treated in this study in which the following response rates were observed: etoposide-cisplatin 12%, taxol-cisplatin 26%, taxol-cisplatin G-CSF 31%. There was no significant difference in the response rates for the two taxol regimens, but the response rates were significantly higher for each taxol regimen compared to etoposide-cisplatin (p < 0.001). The survival results for etoposide-cisplatin versus taxol-cisplatin versus taxol-cisplatin G-CSF are summarized in table 5. Survival for both taxol regimens was superior to etoposide-cisplatin.

Table 5. Survival results

Regimen	Median survival months	One-year survival, %
Etoposide-cisplatin	7.4	31.3
Taxol-cisplatin-filgrastim	10.1	40.4
Taxol-cisplatin	96.6	37.3

Statistical analyses showed significant differences using Wilcoxon analyses with borderline significant differences using log-rank statistics. These results suggest that taxol-cisplatin had a greater effect on earlier survival compared to later survival, an observation which has also been made in one of the meta-analyses involving treatment with combination chemotherapy versus supportive care in advanced NSCLC [24].

The other randomized taxol trial was conducted by European investigators who compared cisplatin 80 mg/m^2 plus taxol 175 mg/m^2 as a 3-hour infusion 80 mg/m^2 plus teniposide 100 mg/m^2 on days 1, 3 and 5 [55]. The investigators enrolled 251 patients in this trial Preliminary results were reported at the Annual Meeting of the American Society of Clinical Oncology in May 1996.

Like the ECOG trial, the response rate on taxol-cisplatin was reported to be higher than that observed with teniposide-cisplatin. However, the preliminary analysis showed no significant difference in survival

So now there are two new regimens which are associated with superior survival compared to older cisplatin-based regimens. Are the taxol-cisplatin and vinorelbine-cisplatin regimens equivalent with respect to survival? Is one better tolerated than the other? The patient characteristics for the French [34] and ECOG trials [54] are summarized in table 6, and the response rates and survival are shown in table 7. The response rates for vinorelbine-cisplatin and for the taxol-cisplatin regimens are approximately 30% and the median survival durations are very similar, namely 10 months. In the French trial, 40% of the patients had stage III disease [33] compared to 19% stage III patients in the ECOG trial [52]. The higher percentage of stage III patients would be expected to have a positive effect on survival However, the French taxol study included patients with poor performance status and there were also fewer female subjects. Both of these factors would be expected to be associated with worse survival Obviously a randomized trial is required to determine whether one of these new regimens is superior, and in fact SWOG is currently conducting a randomized trial comparing taxol-carboplatin to vinorelbine-cisplatin.

Table 6. Phase III trials of new combination regimens – patient characteristics

Investigators	Regimen	Patients	PS2 %	Stage III %	Weight loss >5, %	Females %
LeChevalier et al. [34]	Vinorelbine-cisplatin	206	20	40	NA	11
	Vindesine-cisplatin	200	16	35	NA	10
Bonomi et al. [54]	Etoposide-cisplatin	194	0	19	26	35
	Taxol-cisplatin	187	0	19	25	38
	Taxol-cisplatin-filgrastim	190	0	19	28	37

PS2 = ECOG Performance Status 2.

Future Directions

With the availability of five new drugs which have produced a response rate $\geq 20\%$ in phase II trials and with two of them, vinorelbine and taxol, being associated with superior survival in phase III trials, clinical investigators are faced with a variety of questions. Which two drug regimens should be tested in phase III studies? Should cisplatin and/or carboplatin continue to be included in combination regimens involving new drugs? Which three-drug combinations should be tested? Certainly there are no easy answers to these questions, but hopefully the decision to study a particular new drug combination will be based on the principles of combination chemotherapy which were described earlier [1]. It would be even better if the design of a new regimen were based on laboratory data which showed evidence of synergy for the agents selected for combination.

The carboplatin-taxol regimen has been one of the more extensively studied new two-drug combinations [54–57]. Neurotoxicity manifested as peripheral neuropathy is one of the major toxicities of the cisplatin-taxol regimen [52].

Substituting carboplatin for cisplatin has the potential advantage of reducing neurotoxicity. In three of the carboplatin-taxol studies, taxol was given as a 24-hour infusion [54–56], and in one it was infused over 3 h [57]. The carboplatin dose was based on the Calvert formula, and the area under the

Table 7. Phase III trials of new combination regimens

Investigators	Regimen	Patients	Response %	p	Mean survival/ % yr	p
LeChevalier et al. [34]	Vinorelbine-cisplatin	206	30	0.02	40 weeks/–	0.04
	Vindesine-cisplatin	200	19		31 weeks/–	
Bonomi et al. [54]	Etoposide-cisplatin	194	12		7.4 months/31%	**
	Taxol-cisplatin	187	31	0.001*	9.6 months/37%	
	Taxol-cisplatin- filgrastim	190	26		10.1 months/40%	

* This is the p value for comparison of response rates for etoposide-cisplatin versus the taxol regimens.

**	Wilcoxon test	log rank
Etoposide-cisplatin vs. taxol-cisplatin	p=0.06	p=0.08
Etoposide-cisplatin vs. taxol-cisplatin-filgrastim	p=0.02	p=0.06
Taxol-cisplatin vs. Taxol-cisplatin-filgrastim	p=0.546	p=0.891

curve (AUC) was 6–7 mg/ml/min in the four trials [58]. The maximum recommended dose of taxol was 175–225 mg/m^2. Response rates were >40% in three of the trials [56, 57, 59], while a response rate of 27% was observed in one study [58]. The relatively high response rates, the decrease in neurotoxicity, and the relative ease of administering carboplatin and taxol as a 3-hour infusion on an outpatient basis have attracted the interest of three cooperative groups: The design of the current ECOG trial is cisplatin-taxol (given as a 24-hour infusion) versus carboplatin (AUC of 6)-taxol (225 mg/m^2 i.v. over 3 h) versus cisplatin-taxotere versus cisplatin-gemcitabine. SWOG investigators are currently evaluating vinorelbine-cisplatin versus taxol-carboplatin, and CALGB plans to compare carboplatin-taxol to taxol alone.

In addition to testing the new drugs in stage IV patients, clinical investigators are trying to determine how to incorporate the new drugs into combined modality regimens for the treatment of stage III patients. There is increasing evidence that chemotherapy has a greater effect on survival in stage III patients than in stage IV [60–62] and in one of the trials comparing radiotherapy alone to sequential chemoradiotherapy, long-term survival was improved in patients treated with chemoradiotherapy [63]. It seems likely that the modest prolongation of survival observed with taxol-cisplatin and vinorelbine-cisplatin regimens alone will translate into more substantial improvement in survival when combined with local therapy in the treatment of stage III patients.

References

1   Devita VT Jr: Principles of chemotherapy; in DeVita VT Jr, Hellman S, Rosenberg SA (eds): Cancer: Principles and Practice of Oncology, ed 4. Philadelphia, Lippincott, 1993, pp 276–292.
2   Bitran JD, Desser RK, Messer TR, et al: Cyclophosphamide, Adriamycin, methotrexate, and procarbazine (CAMP) – Effective four drug combination chemotherapy for metastatic non-oat cell bronchogenic carcinoma. Cancer Treat Rep 1976;60:225.
3   Chahinian AP, Arnold DJ, Cohen JM et al: Chemotherapy for bronchogenic carcinoma. JAMA 1977;237:239.
4   Vogl SE, Mehta CT, Cohen MH: MACC: Chemotherapy for adenocarcinoma and epidermoid carcinoma of the lung. Cancer 1979;44:864.
5   Ruckdeschel JC, Finkelstein DM, Ettinger DS, et al: A randomized trial of the four most active regimens for metastatic non-small cell lung cancer. J Clin Oncol 1986;4:14.
6   Eagan RT, Ingle JN, Frytak S, et al: Platinum-based polychemotherapy versus clear hydrogalactitol in advanced non-small cell lung cancer. Cancer Treat Rep 1977;61:1339.
7   Ruckdeschel JC, Finkelstein DM, et al: Chemotherapy for metastatic non-small cell bronchogenic carcinoma: EST: 2575, generation V–A randomized comparison of four cisplatin-containing regimens. J Clin Oncol 1985;3:72.
8   Finkelstein DM, Ettinger DS, Ruckdeschel JC: Long-term survivors in metastatic non-small cell lung cancer: An Eastern Cooperative Group Study. J Clin Oncol 1986;4:702.
9   Bonomi PD, Finkelstein DM, Ruckdeschel JC, et al: Combination chemotherapy versus single agents followed by combination chemotherapy in stage IV non-small cell lung cancer: A study of the Eastern Cooperative Oncology Group. J Clin Oncol 1989;17:1602.
10 Einhorn LH, Loehrer PJ, Williams SD, et al: Random prospective study of vindesine versus vindesine plus cisplatin, plus mitomycin in advanced non-small cell lung cancer. J Clin Oncol 1986;4:1037.
11 Miller TP, Chen TT, Cottman CA, et al: Effects of alternating combination chemotherapy or survival of ambulatory patients with metastatic large-cell and adenocarcinoma of the lung: A Southwest Oncology Group Study. J Clin Oncol 1986;4:502.
12 Gralla RJ, Casper ES, Kelson DP, et al: Cisplatin and vindesine combination chemotherapy for adrenal carcinoma of the lung: A randomized trial investigating two dose schedules, Ann Intern Med 1981;85:414–420.
13 Klastersky J, Sculier JO, Ravez P, et al: A randomized study comparing a high and a standard dose of cisplatin in combination with etoposide in the treatment of advanced non-small cell lung cancer. J Clin Oncol 1986;4:1780.
14 Gandara DR, Crowley J, Livingston RB, et al: Evaluation of cisplatin intensity in metastatic non-small cell lung cancer: A phase III study of the Southwest Oncology Group. J Clin Oncol 1931;11:873.
15 Kreisman H, Ginsberg S, Propert K, et al: Carboplatin or iproplatin in advanced non-small cell lung cancer: A cancer and leukemia group study. Cancer Treat Rep 1989;71:1049.
16 Klastersky J, Sculier JP, Lacroix H, et al: A randomized study comparing cisplatin or carboplatin with etoposide in patients with advanced non-small cell lung cancer: European Organization for Research and Treatment of Cancer. Protocol 07861. J Clin Oncol 1990;8:1556.
17 Cormier Y, Bergeron D, LaForge J, et al: Benefits of polychemotherapy in advanced non-small cell bronchogenic carcinoma. Cancer 1982;50:845.
18 Williams CJ, Woods R, Levi J, et al: Chemotherapy for non-small cell lung cancer: A randomized trial of cisplatin/vindesine versus no chemotherapy. Semin Oncol 1988;15(suppl 7):58.
19 Rapp E, Porter JL, William A, et al: Chemotherapy can prolong survival in patients with advanced non-small cell lung cancer: Report of a Canadian multicenter randomized trial J Clin Oncol 1988;6:633.

20   Cellerino R, Tummarello D, Guidi F, et al: A randomized trial of alternating chemotherapy versus best supportive care in advanced non-small cell lung cancer. J Clin Oncol 1991;9:1453.

21   Kaasa S, Lund E, Thorud E, et al: Symptomatic treatment versus combination chemotherapy for patients with extensive non-small cell lung cancer. Cancer 1988;67:2443.

22   Cartei G, Cartei F, Cantene A, et al: Cisplatin-cyclophosphamide-mitomycin combination chemotherapy with supportive care versus supportive care alone for treatment of metastatic non-small cell lung cancer. J Natl Cancer Inst 1993;85:794.

23   Stewart LA, Pignon JP, Parmar MKB, et al: A meta-analysis using individual patient data randomized clinical trials of chemotherapy in non-small cell lung cancer survival in the supportive care setting. Proc Am Soc Clin Oncol 1994;13:337.

24   Albain KS, Crowley JJ, LeBlanc MN, et al: Survival determinants in extensive non-small cell lung cancer: The Southwest Oncology Group Experience. J Clin Oncol 1991;9:1617.

25   Grilli R, Oxman AD, Julian JA: Chemotherapy for advanced non-small cell lung cancer: How much benefit is enough? J Clin Oncol 1993;11:866.

26   Raby B, Pater J, Mackillop W: A survey of Canadian physicians' belief about the use of radiation and chemotherapy in non-small cell lung cancer. Proc Am Soc Clin Oncol 1994;13:327.

27   Sorensen JB, Hansen HH, Dambernowsky P, et al: Chemotherapy for adenocarcinoma of the lung (WHO III)L: A randomized study of vindesine and methotrexate versus all four days. J Clin Oncol 1987;5:1169.

28   Klastersky J, Sculier JP, Bureau G, et al: Cisplatin versus cisplatin plus etoposide in the treatment of advanced non-small cell lung cancer. J Clin Oncol 1989;7:1087.

29   Rosso R, Salvati F, Ardizzoni A, et al: Etoposide versus etoposide plus high-dose cisplatin in the management of advanced non-small cell lung cancer. Cancer 1990;66:130.

30   Kawahara M, Furuse K, Nagohesa K, et al: A randomized study of cisplatin versus cisplatin plus vindesine for non-small cell lung cancer. Cancer 1991;68:74.

31   Veeder MH, Jett JR, Su JO, et al: A phase III trial of mitomycin C alone versus mitomycin C, vinblastine and cisplatin for metastatic squamous cell lung cancer. Cancer 1992;70:2281.

32   DePierre A, LeBeau B, Chasting C, et al: Results of a phase III randomized study of vinorelbine-cisplatin in non-small cell lung cancer. Proc Am Soc Clin Oncol 1993;12:340.

33   Giaccone G, Splinter T, Fester K, et al: Cisplatin combined with teniposide improves response and survival over VM26 alone in non-small cell lung cancer: A randomized trial of the EORTC Lung Cancer Cooperative Group. Proc Am Soc Clin Oncol 1993;12:331.

34   LeChevalier T, Brisgand D, Doulliard JY, et al: Randomized study of vinorelbine and cisplatin versus vinorelbine alone in advanced non-small cell lung cancer: Results of a European multicenter trial including 612 patients. J Clin Oncol 1994;12:360.

35   Wozniak AJ, Crowley JJ, Balcerzak SP, et al: Randomized phase III trial of cisplatin versus cisplatin plus navelbine in treatment of advanced non-small cell lung cancer: Report of a Southwest Oncology Group Study. Proc Am Soc Clin Oncol 1996;15:374.

36   DePierre A, Lamarie F, Dabouis G, et al: A phase II study of navelbine in the treatment of non-small cell lung cancer. Am J Clin Oncol 1991;14:115.

37   Yokoyama A, Furuse K, Niitani H, et al: Multi-institutional phase II study of navelbine (vinorelbine) in non-small cell lung cancer (abstract). Proc Am Soc Clin Oncol 1992;11:287.

38   Crawford J, O'Rouke M, Schiller JH, et al: Randomized trial of vinorelbine compared with fluorouracil and leucovorin in patients with stage IV non-small cell lung cancer. J Clin Oncol 1996;14:2774.

39   Vokes EE, Rosenberg RK, Johanzeb M, et al: Multicenter phase II study of weekly oral vinorelbine for stage IV non-small cell lung cancer. J Clin Oncol 1955;13:637–644.

40   Chang AY, Kim K, Glick J, et al: Phase II study of taxol, membrane and piroxantrone in stage IV non-small cell lung cancer. J Natl Cancer Inst 1932;85:384.

41   Murphy WK, Fussell AFV, Winn RJ, et al: Phase II study of taxol in patients with untreated advanced non-small cell lung cancer. J Natl Cancer Inst 1993;85:384.

42   Millward MJ, Bishop JF, Friedlander M, et al: Phase II trial of a 3-hour infusion of paclitaxel in previously untreated patients with advanced non-small cell lung cancer. J Clin Oncol 1996;14:142–148.

43   Hainsworth JD, Thompson DS, Greco FA: Paclitaxel by 1-hour infusion: an active drug in metastatic non-small cell lung cancer. J Clin Oncol 1995;13:1609–1614.

44 Francis PA, Rigas JR, Kris MG, et al: Phase II trial of docetaxel in patients with stage III and IV non-small cell lung cancer. J Clin Oncol 1994;112:1232.

45 Fossella FV, Lee JS, Murphy WK, et al: Phase II study of docetaxel for recurrent or metastatic non-small cell lung cancer. J Clin Oncol 1994;12:1238.

46 Fossella FV, Lee JS, Shin JD, et al: Taxotere (docetaxel), an active agent for plantinum refractory non-small cell lung cancer. Preliminary report of a phase II study. Proc Am Soc Clin Oncol 1994;13:336.

47 Rowinsky EK, Ettinger DS: Drug development and new drugs for lung cancer; in Dass HI, Mitchell JB, Johnson DH, Turrisi AT (eds): Lung Cancer: Principles and Practice. Philadelphia, Lippincott-Raven, 1996.

48 Fukuoka M, Niitani H, Suzuki A, et al: A phase II study of CPT-11, a new derivative of camptothecin for previously untreated non-small cell lung cancer. J Clin Oncol 1992;10:16.

49 Masuda N, Fukuoka M, Takada M, et al: CPT-11 in combination with cisplatin for advanced non-small cell lung cancer. J Clin Oncol 1992;10:1775.

50 Anderson H, Lund B, Bach F, et al: Single agent activity of weekly gemcitabine in advanced non-small cell lung cancer: A phase II study. J Clin Oncol 1994.

51 Negoro S, Fukuoka M, Kurita Y, et al: Results of phase II studies of gemcitabine in non-small cell lung cancer (abstract). Proc Am Soc Clin Oncol 1994;13:367.

52 Abratt R, Bezwoda W, Falkson G, et al: Efficacy and safety of gemcitabine in non-small cell lung cancer: A phase II study. J Clin Oncol 1994;12:1535.

53 LeChevalier T, Gottfried M, Gatzemier M, et al: Confirmatory activity of gemcitabine in non-small cell lung cancer. Eur J Cancer 1993;29(suppl 6):160.

54 Bonomi P, Kim K, Chang A, et al: Phase III trial comparing etoposide, cisplatin versus taxol with cisplatin G-CSF versus taxol-cisplatin in advanced non-small cell lung cancer: An Eastern Cooperative Oncology Group trial Proc Am Soc Clin Oncol 1996;15:382.

55 Giaccone G, Splinter T, Postmus P, et al: Paclitaxel-cisplatin versus teniposide-cisplatin in advanced non-small cell lung cancer. Proc Am Soc Clin Oncol 1996;15:373.

56 Langer CJ, Leighton JC, Comis RL, et al: Paclitaxel and carboplatin in combination in the treatment of advanced non-small cell lung cancer: A phase II toxicity response, and survival analysis. J Clin Oncol 1995;13:1860–1870.

57 Belani CP, Aisner J, Hiponia D, et al: Paclitaxel and carboplatin with and without filgrastim support in patients with metastatic non-small cell lung cancer. Semin Oncol 1995;22(suppl 9):7–12.

58 Johnson DH, Paul DM, Hande KR, et al: Paclitaxel plus carboplatin for advanced lung cancer: Preliminary results of a Vanderbilt University phase II trial – Lun-46. Semin Oncol 1995;22(suppl 9): 30–33.

59 Muggia FM, Vafai D, Natale R, et al: Paclitaxel 3-hour infusion given alone and combined with carboplatin: Preliminary results of dose escalation trials. Semin Oncol 1995;22(suppl 9):63–66.

60 Dillman RO, Seagren SL, Propert KJ, et al: A randomized trial of induction chemotherapy plus high-dose radiation versus radiation alone in stage III non-small cell lung cancer. N Engl J Med 1990;323:940–945.

61 LeChevalier T, Arriagada R, Quoix E, et al: Radiotherapy alone versus combined chemotherapy and radiotherapy in non-resectable non-small cell lung cancer: First analysis of a randomized trial in 353 patients. J Natl Cancer Inst 1991;83:417–423.

62 Sause WT, Scott C, Taylor S, Johnson D, Livingstone R, Komaki R, Emami B, Carran WJ, Byhardt RW, Turrisi AT: Radiation Therapy Oncology Group 88-08 and Eastern Oncology Groups 4588L: Preliminary results of a phase III trial in regionally advanced, unrespectable non-small-cell lung cancer. J Natl Cancer Inst 1995;87(3):198–205.

63 Dillman RO, Hernden J, Seagren SL, et al: Improved survival in stage III non-small cell lung cancer: Seven-year follow-up of cancer and leukemia. Group B (CALGB) 8433 Trial J Natl Cancer Inst 1996;88:1210–1215.

Philip D. Bonomi, MD, Rush-Presbyterian-St. Luke's Medical Center,
Professional Building, Suite 809, Section of Oncology,
1725 West Harrison Street, Chicago, IL 60612 (USA)

Schiller JH (ed): Updates in Advances in Lung Cancer. Prog Respir Res.
Basel, Karger, 1997, vol 29, pp 73–90

Chapter 5

••••••••••••••••••••••••

The Emerging Role of Paclitaxel and Carboplatin in Non-Small Cell Lung Carcinoma

Corey J. Langer

Fox Chase Cancer Center, Thoracic Oncology, Philadelphia, Pa., USA

Background

More than 170,000 individuals are diagnosed with lung cancer each year in the US [1]. 75–80% of these cases are histologically designated non-small cell lung cancer (NSCLC). More than 35–40% of all patients with NSCLC have metastatic or stage IV disease at diagnosis [2], an annual incidence of roughly 50,000. As a public health risk, this number is more than double the total incidence of ovarian carcinoma in the US each year. The median survival time for patients with advanced NSCLC who do not receive therapy is extremely poor: 4–6 months at best [3, 4]. Single agents have traditionally yielded response rates of 15–20% at best, while standard combinations through the early 1990s, such as etoposide and cisplatin or mitomycin and vinblastine or vindesine, have yielded response rates of 20–30% with little impact on survival [5–9].

However, after a therapeutic drought of nearly 20 years, a number of new agents with promising activity have emerged, including the taxanes (paclitaxel, docetaxol), topoisomerase I inhibitors (CPT-11, topotecan), gemcitabine and vinorelbinc.

Paclitaxel Single-Agent Studies in Advanced NSCLC

Phase I studies of paclitaxel suggested activity in advanced NSCLC [10, 11]. In 1992, the Eastern Cooperative Oncology Group (ECOG) reported

Table 1. Taxol-NSCLC

Center/country (1st author)	Duration, h	Dose	Stage	OR, %	n (n = 294)	MS, months	1 YS, %
ECOG (Chang) [12]	24	250	III/IV	21	24		40
MDA (Murphy) [13]	24	200	IV	24	25		38.5
Germany (Gatzemeier) [16]	3	225	III/IV	24	50	10	
AEMC (Tester) [17]	3	200	IV	30	20	6.5	
Japan (Sekine) [18]	3	210	III/IV	38	60	11.2	48
Spain (Albergola) [19]	3	210	III/IV	36	62		
Hainesworth [20]	1	200	III/IV	33	36		
	1	135	IV	12	17		

on a randomized phase II study of paclitaxel, merbarone, and piroxantrone in advanced NSCLC [12], in which paclitaxel proved to be the most active single agent in ECOG studies in advanced NSCLC, with a 24% response rate and 1-year survival rate of 41%. Although the survival rates were not formally compared, there was a clear trend toward improvement for patients on the paclitaxel arm, compared to those receiving the other two agents, both of which proved to be inactive and which yielded 1-year survival rates of 23 and 22% respectively. Murphy et al. [13] at M.D. Anderson confirmed the activity of paclitaxel (24-hour infusion); response and 1-year survival rates were similar. With data beginning to emerge suggesting therapeutic equivalence of 3-hour infusion to 24-hour infusion in breast and ovarian carcinoma [14, 15], a number of investigators both in North America and abroad evaluated short, outpatient infusion of paclitaxel in advanced NSCLC [16–19]. Response rates ranged from 22 to 38%, the latter observed in a Japanese study of patients with both stage III and IV disease. Myelosuppression was less pronounced, but nonhematologic toxicities, in particular myalgias/arthralgias and peripheral sensory neuropathy, proved more noticeable. Long-term survival data, though not uniformly available, appear to be comparable to that reported in 24-hour infusion studies (table 1).

More recently, Greco and Hainesworth [20, 21] at Sarah Cannon Cancer Center in Tennessee, have evaluated the efficacy of 1-hour paclitaxel in advanced NSCLC. The pharmacokinetics of 1-hour infusion are virtually identical to those of 3-hour infusion. Like 3-hour infusion, it is markedly less myelosuppressive compared to 24-hour infusion. It has also proven convenient, less costly and safe with few episodes of allergic reactions, despite shortened infusion. Their work suggests a dose-response effect. The response rate for patients receiving 135 mg/m^2 was 12%; at 200 mg/m^2, once every 3 weeks or on a daily fractionated basis for 3 days every 3 weeks, the response rate was considerably higher (36%). At the higher dose, activity was demonstrated in both chemonaive and previously treated patients. Six of 20 chemonaive patients (30%) and 6 of 16 patients with previous platinol exposure (38%) responded to therapy.

Carboplatin in Advanced NSCLC

Carboplatin has a much more favorable toxicity profile compared to cisplatin; it results in considerably less renal toxicity, ototoxicity and neurotoxicity. In the mid-1980s, the ECOG mounted a five-arm phase III randomized trial [22] of cisplatin combinations and cisplatin analogs (EST-1583). Initial therapy with carboplatin, despite producing one of the lowest response rates (9%), led to the best median survival time (31.4 vs. 25.8 weeks or less for each of the other arms). In addition, the incidence of grade 4 toxicity was 3% compared to 13% or higher for each of the other arms. Combination mitomycin, vinblastine and cisplatin, while producing a higher response rate (22%), resulted in the poorest median survival (22.6 weeks). The Cancer and Leukemia Group B (CALGB) in a phase II study of patients with stage III and IV NSCLC administered carboplatin at a dose of 400 mg/m^2 and realized a somewhat higher response rate of 16% compared to the ECOG trial [23]. Finally, a European trial demonstrated equivalent median survival for carboplatin-etoposide compared to the 'standard' cisplatin-etoposide regimen, despite a slightly lower response rate [24]. The carboplatin-based regimen also resulted in less toxicity. In the palliative disease setting, toxicity and survival may be better arbiters of efficacy than response rate.

Combined Paclitaxel (24-Hour Infusion) and Carboplatin in Advanced NSCLC (table 2)

Fox Chase Cancer Center (FCCC)

In June 1993, the FCCC and its network affiliates launched a phase II trial of paclitaxel, in combination with carboplatin, in patients with chemotherapy-naive, histologically proven, advanced NSCLC, stage IV or stage III-B with malignant pleural effusion [25]. Entry criteria stipulated measurable disease, preserved ECOG performance status of 0–1, adequate hematologic hepatic, and renal indices, adequate cardiac function, and no active arrhythmias or congestive heart failure. Patients with brain metastases were allowed on study

Table 2. Taxol-CBDCA combinations: NSCLC

Center (1st author)	Phase	Taxol[1]	Duration, h	Interval, weeks	CBDCA[1]	G-CSF	n (at MTD)	RR (%) at MTD	MS, weeks	1 YS, %
FCCC (Langer)	II	135→175 →215[2]	24	3	AUC7.5	(+)	54	62	54	54
Vanderbilt (Johnson)	I/II	135→175	24	4	300/m² →AUC6	(−)	51 (23)	39	38	32
Maryland (Belani)	I/II	135→175 →200	24	3	AUC5→7→9 →11	(−/+)	30 (16)	56		

CBDCA = Carboplatin; G-CSF = granulocyte-colony stimulating factor; n = total number; (at MTD) = number treated at MTD; MTD = maximally tolerated dose; RR = response rate; MS = median survival; 1 YS = 1-year survival rate.

[1] Maximal tolerated doses italicized.

[2] Intra-patient dose escalation (other studies escalated doses across separate patient cohorts).

if their metastases had been adequately treated and controlled and their performance status was not compromised. Patients with invasive malignancy in the antecedent 3 years or radiation to 30% or more of marrow-bearing bone were excluded from study, as were those who completed radiation or biologic therapy within 4 weeks of study initiation, and those with mixed small cell and non-small cell histology.

Patients received paclitaxel at 135 mg/m² by 24-hour infusion on day 1 in combination with carboplatin dosed to a targeted AUC (area under the concentration-time curve) of 7.5, an approach shown to be more physiologic and less myelosuppressive than conventional BSA dosing [26]. Granulocyte-colony stimulating factor (G-CSF) 5 μg/kg subcutaneously daily for 2 weeks was introduced during the second and subsequent cycles immediately after each chemotherapeutic administration. In the absence of grade 4 neutropenia or grade 3 thrombocytopenia, paclitaxel only was escalated 40 mg/m²/cycle to a maximum dose of 215 mg/m² by cycle 3, the dose range used in the single-agent 24-hour infusion studies at M.D. Anderson and in the ECOG [12, 13]. 25% dose reductions were employed for neutropenic fever or grade 4 thrombocytopenia lasting ≥ 5 days. The carboplatin dose was otherwise fixed. Treatment was repeated at 3-week intervals for a total of 6 cycles.

This was a multi-institutional, phase II trial. 54 patients were enrolled; 43% were treated at FCCC community network affiliates. 53 were evaluable for response and toxicity; 1 patient died on day 10 during the first cycle from acute pulmonary hemorrhage and secondary asphyxia after experiencing an initial marked improvement in baseline dyspnea. The median age was 62 (range 34–84). 69% were male. The majority were symptomatic: 80% had ECOG PS-I. 65% had adenocarcinoma, 22% squamous cell carcinoma, 7% large cell, 4% adenosquamous carcinoma, and 1 NSCLC, not otherwise specified. All but 4 patients (93%) had stage IV or recurrent disease. 52% had lost ≥5% of their baseline weight. 35% had received prior radiation, primarily to the brain or to the thorax.

The potential toxicity of this regimen in advanced NSCLC was a major concern. The goal of this study was to combine carboplatin and paclitaxel at or near their individual maximal tolerated doses. The first cycle doses had previously been piloted in ovarian cancer [27, 28] under the aegis of the Gynecologic Oncology Group (GOG) in a population that was substantially younger with a higher percentage of nonsmokers and therefore fewer smoking related comorbidities. In the NSCLC study, the collaborators sought to escalate the paclitaxel dose further, which would potentially heighten myelosuppression. Fortunately, hematologic toxicity, though substantial, was manageable: there was a 71% incidence of grade ≥3 neutropenia, 45% grade 4; 48% incidence of grade ≥3 thrombocytopenia, 23% grade 4, and 34% incidence of grade ≥3 anemia, only 2% grade 4. During the first cycle in the absence of G-CSF, granulocytopenia was dose-limiting, resulting in a 57% incidence of grade ≥3 neutropenia. With the introduction of G-CSF during the second and subsequent cycles, this incidence dropped to 35% during cycle 2, and then decreased to 22% or less consistently during subsequent cycles. More importantly, from a clinical perspective, all 7 episodes of neutropenic fever during this trial occurred during the first cycle.

Not surprisingly, thrombocytopenia and anemia were not affected by G-CSF. The cumulative incidence of thrombocytopenia, all grades, rose from 6% during the second and third cycles to 30% by cycles 5 and 6. Anemia, similarly, was cumulative. Both anemia and thrombocytopenia were more pronounced in patients who had received prior radiation, suggesting an apparent abscopal effect. Grade 3 nonhematologic toxicities included anorexia in 23% of patients, fatigue in 21%, nausea and vomiting in 8%, myalgias and arthralgias in 6%, and peripheral sensory neuropathy and hemorrhagic cystitis in 2% each. Neuropathy, all grades, was cumulative, rising from a 2% incidence during the second cycle to 38% by cycle 6. Myalgias and arthralgias, likewise, proved cumulative. There were no hypersensitivity reactions.

The median first cycle carboplatin dose was 429 mg/m^2, with a wide range: 275–709 mg/m^2. Treatment delays were relatively infrequent, but clearly cumulative, rising from 10% during the second cycle to 30% or more during the fifth and sixth cycles. More than 50% of the patients achieved full-dose paclitaxel by cycle 3; by cycle 5, more than 70% received paclitaxel at full dose. Fewer than 20% of patients remained at the original dose during the third and subsequent cycles.

The overall objective response rate was 62% (confidence intervals 48–74%). There were 5 CT-documented complete responses (CR) lasting 31–57 weeks, and 28 partial responses (PR) ranging from 4 to 148 weeks. Five patients had minor regression, 10 had stable disease, and 5 progressive disease. If patients with minor response are grouped with those who achieved CR or PR, 55% achieved response within the first 6 weeks of treatment, that is before they had an opportunity to have their paclitaxel dose escalated to 215 mg/m^2, therefore raising the question of whether dose intensification of 24-hour paclitaxel above 175 mg/m^2 is necessary.

At a median potential follow-up of 34 months and a minimum follow-up of 30 months, the duration of response ranges from 1 to 34+ months and the median response duration is 6 months. Median survival is 12.5 months, 1-year survival rate 54%, and 2-year survival rate 15% [29].

Vanderbilt

Johnson et al. [30] from Vanderbilt have mounted a phase I/II study of the same regimen. Using conventional BSA dosing for carboplatin (300 mg/m^2), in combination with paclitaxel 135 mg/m^2 over 24 h, myelosuppression was treatment-limiting: leukopenia was pronounced with median WBC nadir of 1.6/μl (range 0.8–4.3), and thrombocytopenia was unpredictable with nadirs as low as 7,000/μl. With conversion to AUC-based dosing (targeted AUC of 6) in a subsequent cohort of patients, median ANC nadirs rose to 3.8/μl and thrombocytopenia proved more predictable; the lowest observed platelet nadir was 60,000/μl. A third cohort was able to tolerate paclitaxel dose escalation to 175 mg/m^2, with no exacerbation of myelosuppression. Among 51 evaluable patients, there were 14 PR for an overall response rate of 27%, 95% confidence intervals: 17–41%. At the maximally tolerated doses, the response rate was 39%. The overall median survival was 38 weeks, and overall 1-year survival rate 32% [31]. The median survival for those with PS-1 was 41.3 weeks and for those with PS-0 it was 50.4 weeks. This study clearly demonstrated the utility of AUC-based dosing both in reducing myelosuppression and in enhancing tolerance.

University of Maryland

Belani et al. [32] at the University of Maryland combined paclitaxel 135 mg/m^2 24-hour infusion with carboplatin (Calvert formula) in escalating doses in 3- to 5-patient cohorts. Once the maximal carboplatin dose was established, the paclitaxel dose was escalated in sequential cohorts. Myelosuppression was

Table 3. Taxol-CBDCA combinations: NSCLC

Center (1st author)	Phase	Taxol[1]	Dura-tion, h	Inter-val, wks	CBDCA[1]	G-CSF	n at MTD	RR (%) at MTD	MS, weeks	1 YS, %
USC (Natale)	I/II	150→250 (*225*)	3	3	AUC 6	(−)	51 (28)	61	40+	38 (all)
Colorado (Bunn)	I/II	135→*225*	3	3	250→*400*/m²	(−)	29	50		
Hopkins (Rowinsky)	I/II	175→*225*	3	3	AUC 7→9	(−)	23 (15)	47		
EORTC (Giaccone)	I/II	100–175	3	4	300→*400*/m²	(−)	19	5		
	I/II	≥175–*250*	3	4	300->*400*/m²	(−)	30	20		

[1] Maximal tolerated doses underlined.

the major dose-limiting toxicity. In combination with a carboplatin AUC of 7, in the absence of G-CSF support, the maximal paclitaxel dose was 175 mg/m². Combined with neupogen, the maximum tolerated paclitaxel dose was 200 mg/m². Of 30 evaluable patients at all dose levels, 15 responded, with 1 CR. The response rate for those receiving paclitaxel doses <175 mg/m² was 38%; for those receiving doses of ≥175 mg/m², it was 59%. Long-term survival data have not yet been reported.

Combination Paclitaxel (3-Hour) and Carboplatin in Advanced NSCLC (table 3)

University of Southern California
Vafai and co-workers [33] at the University of Southern California combined a fixed dose of carboplatin (AUC 6) with escalating doses of paclitaxel (3-hour infusion) in sequential cohorts. G-CSF was not used. 49 patients were accrued; median age was 62 (range 46–81) 23 patients had unresectable stage III NSCLC, and 26 stage IV NSCLC. 176 treatment courses were evaluable for toxicity. Myelosuppression was relatively mild: grade 4 leukopenia occurred in just 31% of patients, and there was absolutely no grade 4 thrombocytopenia, in contrast to 24-hour paclitaxel studies. The incidence of myalgias/arthralgias and sensory neuropathy was dose-related with grade 3 events occurring at paclitaxel doses of 200 mg/m² or higher. Grade 3 sensory neuropathy occurred

in 3 of 5 patients treated at 250 mg/m^2 paclitaxel. These symptoms did not generally respond to steroids or nonsteroidal anti-inflammatory agents and often proved cumulative; but at 225 mg/m^2, grade 3 myalgias and arthralgias or grade 3 sensory neuropathy occurred in only 8 of 23 patients. These symptoms stabilized or tended to improve when the dose of paclitaxel was lowered to 200 mg/m^2. Objective responses have been documented in 26 patients for a major response rate of 62%. There have been two CR. The median survival is 10 months.

The lack of significant thrombocytopenia was an unexpected observation, attributed at least in part to the use of the Calvert formula, a rational application of pharmacokinetics to the clinical dosing of carboplatin, which accounts for this agent's renal clearance and predictable myelotoxicity. The carboplatin dose, when recalculated using body surface area, ranged from 240 to 600 mg/m^2.

In addition, in the context of historic studies of carboplatin alone in which thrombocytopenia was either comparable or more pronounced [34], it has been postulated that paclitaxel may actually yield a protective effect on carboplatin-induced megakaryocytic toxicity.

A phase II pharmaceutical-sponsored study evaluating the utility of this combination at the MTDs identified by Natale and co-workers in comparison to standard etoposide-cisplatin has completed accrual and will most likely be reported within the next year.

Johns Hopkins

Rowinsky et al. [35] at Johns Hopkins mounted a phase I study of paclitaxel 3-hour infusion followed by carboplatin in chemotherapy-naive patients with stage IV NSCLC. The starting dose of paclitaxel was 175 mg/m^2; the starting dose of carboplatin was calculated to achieve a targeted AUC of 7 mg/ml/min. The doses of both agents were escalated in sequential cohorts. Treatment was repeated at 3-week intervals. 23 patients were accrued. The median age was 63 (range 44–78). Eleven had ECOG PS-0, 12 PS-1. Nine had received prior RT. Dose-limiting neutropenia and thrombocytopenia were consistently observed at a paclitaxel dose of 225 mg/m^2 and carboplatin AUC of 9. At this dose level, 6 dose-limiting events were experienced by 4 of 6 patients during 5 of 23 total treatment courses. The projected MTDs were paclitaxel 225 mg/m^2 and carboplatin AUC 7. Seven of 19 patients with measurable tumor responded with either PR or CR; 5 additional patients achieved ≤50% reductions in their tumor size. Survival data are not available.

Rowinsky et al. [35] indicated that further dose escalation of paclitaxel above 225 mg/m^2 might be possible with G-CSF support; however, recognizing that progressive neuropathy would very likely preclude further, meaningful dose escalation, this plan was not carried out.

University of Colorado

Bunn et al. [36] at the University of Colorado in a phase I-II study treated patients with advanced, nonresectable stage IIIb and IV NSCLC with paclitaxel given as a 3-hour intervenous infusion followed by 30-min carboplatin infusion. Patients were assigned to 1 of 7 successive treatment groups, with paclitaxel dose increased from 135 to 225 mg/m^2, and carboplatin increased from 250 to 400 mg/m^2. Pharmacokinetic dosing based on the Calvert formula was not used. As of March 1995, 42 patients had been accrued. Median age was 61. 38 of 42 (90%) had stage IV disease. 32 patients had ECOG PS-0 or PS-1; 10 patients PS-2. 24 had adenocarcinoma, 11 squamous cell carcinoma, 2 adenosquamous carcinoma, and 5 large cell undifferentiated tumors. During 118 treatment cycles, there were only 8 episodes of grade 4 neutropenia, none during cycle 1. Because myelosuppression was mild, 3 patients only received G-CSF. There was only 1 episode of grade 4 thrombocytopenia at the highest dose level tested to date (paclitaxel 200 mg/m^2, carboplatin 400 mg/m^2). Nonhematologic toxicities were also mild: grade 3 fatigue and malaise occurred in only 4 patients. Allergic reactions developed in 4 patients prompting protocol discontinuation in each. Neurotoxicity was relatively infrequent, as was grade 3 or 4 nausea and vomiting.

A dose-response relationship appeared to develop. In the two groups receiving the lowest doses, there were no responders. In the highest dose groups, with paclitaxel given at a dose of 175 and 200 mg/m^2 respectively, 6 of 12 evaluable patients (50%) responded (95% confidence interval, 21–79%). It is too early to determine if improvements in response rate in the higher dose groups will translate into improved 1-year survival rate.

EORTC

Giaccone et al. [37] combined paclitaxel at an initial dose of 100 mg/m^2 with a fixed carboplatin dose of 300 mg/m^2, at 4-week intervals in patients who were chemotherapy-naive. At least 2 patients were accrued at each dose level; 62 patients have been entered. Paclitaxel was increased from 100 to 225 mg/m^2. Subsequently, the carboplatin dose was raised to 400 mg/m^2, in conjunction with further escalation of paclitaxel dose to 250 mg/m^2. Hematologic toxicities were minimal to moderate. At paclitaxel doses exceeding 200 mg/m^2, nonhematologic toxicities, specifically bone pain, myalgias and peripheral neurotoxicity, occurred most frequently. In 50 evaluable patients, 6 major responses were observed, 5 at paclitaxel doses of 175 mg/m^2 or higher, again suggesting a dose-response relationship for short infusion paclitaxel in NSCLC [37].

Combination Paclitaxel (1 Hour) and Carboplatin in Advanced NSCLC

Fox Chase Cancer Center

As previously noted, Hainesworth and co-workers [20, 21] have demonstrated activity for paclitaxel administered as a 1-hour infusion at doses of 135–200 mg/m² every 3 weeks in NSCLC. In November 1994, a phase II trial of combination carboplatin (AUC 7.5) and paclitaxel (1 h) was implemented at FCCC [29] with the expectation that the response and survival rates observed in the previous 24-hour paclitaxel study would be maintained without the cost and inconvenience of overnight hospitalization and obligatory use of G-CSF. In the absence of grade 4 myelosuppression, the paclitaxel dose was increased 35 mg/m²/cycle from 175 mg/m² to a maximum dose of 280 mg/m². G-CSF was introduced only for grade 4 neutropenia. As with the original 24-hour infusion study at FCCC, treatment cycles were administered every 3 weeks for a total of 6 cycles. 22 patients with advanced, measurable NSCLC were initially accrued. Eligibility criteria were identical to the original 24-hour study; they mandated stage IV or recurrent disease or stage IIIb disease with malignant pleural effusion. Those with ECOG PS-2 or greater were excluded. The median age was 64 (range 34–80). Cumulative grade 3 peripheral sensory neuropathy occurred in 6 of the first 22 patients enrolled, generally at paclitaxel doses > 210 mg/m², obligating 3 to be removed from this study, in the absence of progressive disease. Consequently, the study was revised. 35 subsequent patients (cohort B) received a lower starting dose of paclitaxel (135 mg/m²) boosted 40 mg/m²/cycle in the absence of grade 4 myelosuppression to a maximum dose of 215 mg/m², the dosing schema originated in FCCC 93-024 (table 4). A comparison of toxicity, response, and survival is shown in table 5. Patients in cohort B experienced considerably less myelosuppression, neurotoxicity and fatigue compared to the original cohort, but overall response rate was only half that observed in either cohort A or the previous 24-hour infusion trial. It is too early to discern if the compromise in response rate will compromise 1-year survival rate.

The results of the 1-hour study suggest that the dose threshold for activity for 1-hour infusion is higher than that observed in 24-hour paclitaxel infusion combinations, and that the higher doses required to observe this level of activity resulted in substantially more neuropathy and more myalgias and arthralgias.

Sarah Cannon Cancer Center

A similar, multi-institutional trial has been mounted by Hainesworth and co-workers [39] combining 1-hour paclitaxel (225 mg/m² i.v. day 1) with carboplatin (AUC 6) every 21 days for a maximum of 8 courses. The eligibility

Table 4. FCCC 94-064 taxol dose escalation

Original (A)	Revised (B)

```
175 ⟍              135 ⟍
      ⟩ 35               ⟩ 40
210 ⟋              175 ⟋
      ⟩ 35               ⟩ 40
245 ⟋              215 ⟋
      ⟩ 35
280 ⟋
```

criteria match those of the FCCC study. As of November 1995, 82 patients were enrolled. 50 were evaluable with a 45% objective response rate. 18% have experienced grade 3 neuropathy; only 6% have sustained grade 3 or 4 arthralgias/myalgias. Although G-CSF has not been used routinely, there have been only 3 episodes of neutropenic fever requiring hospitalization.

Ottawa Regional Cancer Center

Finally, the preliminary results of a phase II trial orchestrated by Evans and co-workers [40] in Canada are available. Paclitaxel at a relatively conservative dose of 175 mg/m^2 was combined with carboplatin at a calculated AUC of 6, based on the Calvert formula; cycles were repeated every 28 days. To date, 4 PR have been observed in 11 evaluable patients. Median neutrophil nadirs were 1.8/µl.

Cooperative Group Trials

The activity of paclitaxel/carboplatin and its relative safety and convenience as an outpatient regimen have made it a comparitor in three separate cooperative group, randomized phase III trials. ECOG 1594 (table 6), which recently opened to accrual, compares this regimen to the new 'standard' combination of paclitaxel 135 mg/m^2 over 24 h and cisplatin 75 mg/m^2 and two other cisplatin combinations: gemcitabine 1 g/m^2 days 1, 8 and 15 and cisplatin

Table 5. A comparative analysis of sequential 24-hour paclitaxel-carboplatin (93-024) and 1-hour paclitaxel-carboplatin (94-064) combinations

	Study (FCCC)		
	93-024	94-064 (A)	94-064 (B)
Carboplatin (AUC)	7.5	7.5	7.5
Paclitaxel, mg/m^2	135–215	175–280	135–215
Paclitaxel – duration, h	24	1	1
Patients, n	54	22	35
Cycles administered	268	98	150
Toxicity, %			
Neutropenia grades 3+4 (4)	71 (45)	82 (73)	66 (34)
Thrombocytopenia grades 3+4 (4)	48 (23)	50 (18)	26 (14)
Myalgias grades 1–3 (3)	36 (6)	*68 (9)*	43 (3)
Neuropathy grades 1–3 (3)	38 (2)	*77 (27)*	51 (6)
Fatigue grades 1–3 (3)	79 (21)	77 (23)	46 (6)
G-CSF cycles/patients, %	*80/100*	31/55	20/26
Major response (CR), %	62 (9)	55 (4)	27 (0)
1-Year survival, %	54	45	46

Italicized figures represent a statistically significant difference for the results obtained.

100 mg/m^2 day 1; and taxotere 75 mg/m^2 day 1, combined with cisplatin 75 mg/m^2 day 1. The Southwest Oncology Group (SWOG) recently demonstrated the superiority of vinorelbine and cisplatin to cisplatin alone; the 1-year survival rates were 35.4 and 16.4% respectively. The newly opened SWOG phase III trial compares the recently identified 'standard' regimen, vinorelbine and cisplatin, to paclitaxel and carboplatin. CALGB has also mounted a phase III randomized trial isolating the contribution of carboplatin to this regimen, comparing paclitaxel alone to paclitaxel and carboplatin. In light of the relatively promising 1-year survival rates of paclitaxel alone [12, 13], it is not clear if the addition of carboplatin truly enhances its efficacy.

These three trials represent state-of-the-art therapy. They should accrue quickly, with data emerging in 2–4 years. Unanswered issues, however, persist. 24-hour infusion, though clearly less convenient and more expensive, may be superior to ≤3-hour infusion. To date, this issue is not being addressed in a randomized fashion in advanced NSCLC. Identification of the precise threshold dose needed for optimal response and survival in short infusion studies has not occurred, but should also be explored. Higher doses of paclitaxel (≥ 200–225 mg/m^2), though possibly more effective, are clearly more neurotoxic.

Table 6. E1594 schema

Randomize:	Arm A: paclitaxel + cisplatin
	Paclitaxel, 135 mg/m^2 i.v. over 24 h, day 1
	Cisplatin, 75 mg/m^2 i.v. over 1 h, day 2
	Each cycle to be repeated every 3 weeks
Randomize:	Arm B: gemcitabine + cisplatin
	Gemcitabine 1,000 mg/m^2 i.v. over 30 min, days 1, 8, 15
	Cisplatin 100 mg/m^2 i.v. over 1 h, day 1
	Each cycle to be repeated every 4 weeks
Randomize:	Arm C: cisplatin + taxotere
	Cisplatin, 75 mg/m^2 i.v. over 1 h, day 1
	Taxotere, 75 mg/m^2 i.v. over 1 h, day 1
	Each cycle to be repeated every 3 weeks
Randomize:	Arm D: paclitaxel + carboplatin
	Paclitaxel, 225 mg/m^2 i.v. over 3 h, day 1
	Carboplatin (AUC 6.0), day 1
	Each cycle to be repeated every 3 weeks

Integration of Paclitaxel/Carboplatin into Regimens for Locally Advanced NSCLC

Data in locally advanced NSCLC have emerged demonstrating the benefits of combining chemotherapy with local measures. Induction chemotherapy followed by definitive radiation compared to radiation alone has demonstrated an improvment in systemic control in three well-conducted phase III trials with enhanced long-term survival rates [41–43]. Radiosensitizing chemotherapy has increased local control rates, which, in three randomized trials from abroad [44–46], has yielded improved long-term survival rates.

The stabilization of the microtubulin polymer by paclitaxel leads to an accumulation of cells in the G-2 and M portions of the cell cycle, when radiation sensitization is maximized [47]. In preclinical cell lines, including lung, ovary, cervix, and astrocytoma, the radiation enhancement ratio of paclitaxel is 50% higher than radiation alone [48, 49]. Choy et al. [49] mounted a phase I trial in patients with locally advanced NSCLC employing radical thoracic RT (60 Gy) in conjunction with weekly paclitaxel administered over 3 h, approximately 3–4 h prior to RT. The initial paclitaxel dose was conservative at 10 mg/m^2 weekly. Doses were escalated by 10 mg/m^2 in sequential cohorts of patients. At 70 mg/m^2 weekly, dose-limiting, grade IV esophagitis requiring hospitalization was observed. However, at a paclitaxel dose of 60 mg/m^2 weekly, only 1 of 7 patients experienced grade 3 esophagitis, prompting the designation of 60 mg/m^2/week as the maximally tolerated dose.

In a phase II study reported at the American Society of Clinical Oncology meeting in 1996, 33 patients, all with stage IIIa and IIIb disease, were treated with concurrent TRT (60 Gy) and paclitaxel 60 mg/m^2 weekly for 6 weeks [50, 51]; 29 proved evaluable. The overall response rate was 76%. Median progression-free survival was 11.1 months. At 18 months, 21% remained free from progression, and over 50% remained alive, making this regimen competitive with other combined modality radiosensitizing approaches in advanced NSCLC. Myelosuppression and neurosensory toxicity were not dose-limiting. The major toxic reactions affected normal tissue in the radiation fields. 37% of the patients experienced grade 3 or 4 esophagitis, 9% experienced grade 3 or 4 pneumonitis.

In 1994, investigators at FCCC and its network affiliates mounted a trial evaluating paclitaxel and carboplatin induction therapy followed by a phase I assessment of concurrent paclitaxel and carboplatin administered at 3-week intervals during radical thoracic RT [52, 53]. The goal of this trial was to capitalize on the systemic benefits of induction therapy and the potential local benefits of concurrent chemoradiation with 'systemic' doses maintained during radiation. Three specific objectives existed: (1) to evaluate the activity and tolerance of induction therapy; (2) to determine the maximum tolerated doses of paclitaxel and carboplatin every 3 weeks during radical RT after induction therapy, and (3) to determine the efficacy of G-CSF priming prior to induction therapy, compared to conventional G-CSF alone.

Eligibility stipulated good prognosis (KPS ≥ 70%; ≤ 5% weight loss) patients with IIIb or bulky IIIa NSCLC. Induction therapy consisted of two cycles of paclitaxel, 175–225 mg/m^2 over 3 h and carboplatin (targeted AUC of 7.5) administered days 1 and 22. G-CSF 5 µ/kg was administered days 2–15 and 23–36 to all patients; half were randomized to priming G-CSF daily for 5 days prior to day 1 chemotherapy. On day 43, radical thoracic radiation (60 Gy/30 Fx/5 days/week) was initiated. FCCC patients only received chemotherapy during RT. The initial carboplatin dose was AUC 3.75 and the initial paclitaxel dose was 67.5 mg/m^2 over 3 h on days 43 and 64. In the absence of dose-limiting toxicity, phase I dose escalation in 3 patient cohorts proceeded to a maximum carboplatin AUC of 5.0 and paclitaxel dose of 175 mg/m^2.

To date, 32 patients have received induction therapy. Median age is 58 (range 41–70); 56% have ECOG PS-1, 53% are male and 78% have IIIb disease. 29 are evaluable for toxicity and 27 for response. One was taken off study because of cerebral vascular accident; another with baseline anemia and possible underlying hematologic disorder died from neutropenic fever during the first cycle. With the exception of this patient, myelosuppression has been mild during induction therapy, prompting an increase in paclitaxel dose to 225 mg/m^2 days 1 and 22 after the first 7 patients were accrued. The phase III portion of the study evaluating G-CSF priming remains coded. 16 patients have received concurrent thoracic radiation and chemotherapy in sequential cohorts with paclitaxel dose days 43 and 64 escalated to 175 mg/m^2. Myelosuppression to date has been mild. During concurrent TRT-CT, 2 of 13 evaluable patients have experienced grade III esophagitis. The severity of esophagitis has generally corresponded to the length of the esophagus in the treatment field: grade 1 in all 6 patients with esophageal exposure < 16 cm; grade ≥ 2 in 6 of 7 patients with length of radiated esophagus > 16 cm. Three episodes of delayed steroid-responsive pulmonary toxicity have occurred 2–6 months after the conclusion of combined modality treatment. There has been no significant neuropathy or myalgias/arthralgias. The major response rate to induction therapy is 38% and to combined modality thaerapy 75%. The 1-year survival rate for the first 21 patients, all of whom have a minimum potential follow-up of 1 year, is 62%. This outpatient approach, combining induction therapy followed by concurrent chemoradiation, is being considered by the Radiation Therapy Oncology Group for a group-wide phase II trial. It is not clear if growth factor is necessary in this regimen. Other investigators are evaluating similar approaches with paclitaxel and carboplatin either alone or in combination with other chemotherapeutics in the setting of RT.

Potential Utility in Resectable NSCLC

The combination of paclitaxel-carboplatin may realize its ultimate fulfillment in earlier stages of NSCLC. The CALGB has mounted a phase III randomized trial of carboplatin/paclitaxel (4 cycles) versus observation in patients with surgically resected T2N0 NSCLC. Other investigators are evaluating this combination as induction therapy prior to surgery in resectable stage III NSCLC. This concept has been evaluated in phase III trials, one in the US and another abroad, using older cisplatin-based therapies. Both trials, though relatively small in accrual, demonstrated striking survival advantages for neoadjuvant chemotherapy followed by surgery, compared to surgery alone in patients with locally advanced, *resectable* NSCLC [54, 55]. Whether the paclitaxel carboplatin combination will prove to be a safer, more efficacious combination in these settings requires active investigation.

References

1   Boring C, Squires T, Tong T, Montgomery S: Cancer statistics, 1994. CA Cancer J Clin 1994;44: 7–26.

2   Minna JD, Pass H, Glatstein E, et al: Cancer of the lung; in DeVita V, Hellman S, Rosenberg SA (ed): Cancer – Principles and Practice of Oncology. Philadelphia, Lippincott, 1989, pp 591–705.

3   Cartei G, Cartei F, Cantone A, et al: Cisplatin-cyclophosphamide-mitomycin combination chemotherapy with supportive care vs. supportive care alone for treatment of metastatic non-small cell lung cancer. J Natl Cancer Inst 1993;85:749–800.

4   Rapp E, Pater JL, Willen A, et al: Chemotherapy can prolong survival in patients with advanced non-small cell lung cancer. Report of a Canadian multicenter randomized trial. J Clin Oncol 1988; 6:633–641.

5   Kris MG, Gralla PJ, Kelsen DP, et al: Trial of vindesine plus mitomycin in stage III (AJCC stage IV) non-small cell lung cancer. An active out-patient regimen. Chest 1985;87:368–372.

6   Ruckdeschel JC, Finkelstein DM, Mason BA, et al: Chemotherapy for metastatic non-small bronchogenic carcinoma: EST 2575, generation V – A randomized comparison of four cisplatin containing regimens. J Clin Oncol 1985;3:72–79.

7   Ruckdeschel JC, Finkelstein DM, Ettinger DS, et al: A randomized trial of the four most active regimens for metastatic non-small cell lung cancer. J Clin Oncol 1986;4:14–22.

8   Finkelstein DM, Ettinger DS, Ruckdeschel JC: Long-term survivors in metastatic non-small cell lung cancer: An Eastern Cooperative Group Study. J Clin Oncol 1986;4:702–709.

9   Comis R, Kris MG, Lee JS, et al: Multicenter randomized trials in 673 patients comparing the combination of edatrexate, mitomycin, and vinblastine (EMV) with mitomycin and vinblastine in patients with stage IV and IIIB non-small cell lung cancer. Lung Cancer 1994;11(suppl 1): 119(abstract).

10 Rowinsky EK, Cazenave LA, Donehowar RC: Taxol: A novel investigational antineoplastic agent. J Natl Cancer Inst 1990;82:1247–1259.

11 Rowinsky EK, Gilbert M, McGuire WP, et al: Sequences of taxol and cisplatin: A phase I and pharmacologic study. J Clin Oncol 1991;9:1692–1703.

12 Chang AY, Kim K, Glick J, et al: Phase II study of taxol, merbarone and piroxantrone in stage IV non-small cell lung cancer: The Eastern Cooperative Oncology Group experience. J Natl Cancer Inst 1993;85:388–394.

13 Murphy WK, Fossella FV, Winn RJ, et al: Phase II study of taxol in patients with untreated advanced non-small cell lung cancer. J Natl Cancer Inst 1993;95:384–388.

14 Swenerton K, Eisenhauer E, ten Bokkel Huinink WW, et al: Taxol in relapsed ovarian cancer: High vs. low and short vs. long infusion: A European-Canadian study coordinated by the NCI Canada Clinical Trials Group (abstract). Proc Am Soc Clin Oncol 1993;12:25.9.

15 Nabholtz JM, Gelmon K, Bontenbal M, et al: Randomized trial of two doses of taxol in metastatic breast cancer: An interim analysis. Proc Am Soc Clin Oncol 1993;12:60(abstr 42).

16 Gatzemeier U, Pawel JV, Heckmayer M, et al: Phase II study with paclitaxel in advanced inoperable non-small cell lung cancer: The European experience. Lung Cancer 1994;11(suppl 2):246.

17 Tester W, Cohn J, Desai A, et al: Phase II study of short infusion paclitaxel in patients with stage IV non-small cell lung cancer. Proc Am Soc Clin Oncol 1995;14(abstr 1189).

18 Sekine I, Nishiwaki Y, Watanabe K, Yoneda S, Saijo N, Kituchi K: Phase II study of 3-hour infusion of paclitaxel in previously untreated non-small cell lung cancer. Clin Cancer Res 1996;2:941–945.

19 Albergola V, Rosell R, Gonzalez-Larriba JL, Molina F, Ayala F, Garcia-Conde J, et al: Single-agent taxol 3-hour infusion, in untreated advanced non-small cell lung cancer. Ann Oncol 1995; 6(suppl 3):49–52.

20 Hainesworth JD, Thompson FD, Greco FA: Paclitaxel by 1-hour infusion: An active drug in metastatic non-small cell lung cancer. J Clin Oncol 1995;13:1609–1614.

21 Hainesworth JD, Greco A: Paclitaxel in lung cancer: 1-hour infusion given alone or in combination chemotherapy. Semin Oncol 1995;22(suppl 15):45–49.

22   Bonomi PD, Finkelstein PM, Ruckdeschel JC, et al: Combination chemotherapy versus single agents followed by combination chemotherapy in stage IV non-small cell lung cancer. A study of the Eastern Cooperative Oncology Group. J Clin Oncol 1989;7:1602–1613.

23   Green MR, Kreisman H, Lickart S, et al: Carboplatin in non-small cell lung cancers: The Cancer and Leukemia Group B experience: Carboplatin (JM-9); in Bunn PA Jr, Canetta R, Ozols R, et al (eds): Current Perspectives and Future Directions. Philadelphia, Saunders, 1990, pp 301–306.

24   Klasterskey J, Sculier JP, Lacroix H, et al: A randomized study comparing cisplatin or carboplatin with etoposide in patients with advanced non-small cell lung cancer: European Organization for Research and Treatment of Cancer Protocol 07861. J Clin Oncol 1990;8:1556–1562.

25   Langer CJ, Leighton JC, Comis RL, et al: Paclitaxel and carboplatin in combination in the treatment of advanced non-small cell lung cancer: A phase II toxicity, response and survival analysis. J Clin Oncol 1995;13:1860–1870.

26   Calvert AH, Newell DR, Gumbrell LA, et al: Carboplatin dosage: Prospective evaluation of a simple formula based on renal function. J Clin Oncol 1989;7:1748–1756.

27   Bookman MA, McGuire WP, Kilpatrick D, et al: Phase I Gynecologic Oncology Group study of 3-hour and 24-hour paclitaxel with carboplatin as initial therapy for advanced epithelial ovarian cancer (abstract). Proc Am Soc Clin Oncol 1995;14:275.

28   Bookman MA, McGuire WP, Kilpatrick D, Keenan E, Hogan WM, Johnson SW, O'Dwyer P, Rowinsky E, Gallion HH, Ozols RF: Carboplatin and paclitaxel in ovarian carcinoma: A phase I study of the Gynecologic Oncology Group. J Clin Oncol 1996;14:1895–1902.

29   Langer C, Rosvold E, Millenson M, Smith M, Kosierowski R, McAleer C, Bonjo C, Ozols R: Paclitaxel by 1- or 24-hour infusion combined with carboplatin in advanced non-small cell lung cancer: A comparative analysis. Proc Am Soc Clin Oncol 1997;16(abstr 1625).

30   Johnson DH, Paul DM, Hande KR, DeVore RF: Paclitaxel plus carboplatin for advanced lung cancer: Preliminary results of a Vanderbilt University phase II trial LUN-46. Semin Oncol 1995; 22(suppl 9):30–33.

31   Johnson DH, Paul DM, Hande KR, Shyr Y, Clarke C, Murphy B, Lewis M, DeVore RF III: Paclitaxel plus carboplatin in advanced non-small cell lung cancer: A phase II trial. J Clin Oncol 1996;14:2054–2060.

32   Belani C, Aisner J, Hiponia D, Engstrom C: Paclitaxel and carboplatin with or without filgastrim support in patients with metastatic non-small cell lung cancer. Semin Oncol 1995;22(suppl 9):7–12.

33   Vafai D, Israel V, Zaretsky S, et al: Phase I/II trial of combination carboplatin and taxon in non-small cell lung cancer (abstract). Proc Am Soc Clin Oncol 1995;14:852.

34   Kearns CM, Belani CL, Erkmen K, et al: Reduced platelet toxicity with combination carboplatin and paclitaxel: Pharmacodynamic modulation of carboplatin-associated thrombocytopenia (abstract). Proc Am Soc Clin Oncol 1995;14:170.

35   Rowinsky EK, Flood WA, Sartorius SE, et al: Phase I study of paclitaxel as a 3-hour infusion followed by carboplatin in untreated patients with stage IV non-small cell lung cancer. Semin Oncol 1995;22(suppl 9):48–54.

36   Bunn PA, Kelly K: A phase I study of carboplatin and paclitaxel in non-small cell lung cancer: A University of Colorado Cancer Center Study. Semin Oncol 1995;22(suppl 9):2–6.

37   Giaccone G, Huizing M, Postmus P, Ten Bokkel Huinink WW, et al: Dose finding and sequencing study of paclitaxel and carboplatin in non-small cell lung cancer. Semin Oncol 1995;22(suppl 9):78–82.

38   Langer C, Kaplan R, Rosvold E, Millenson M, Smith MR, Johnson C, Kosierowski R, McAleer C, Bonjo C, Ozols R: Paclitaxel by 1-hour infusion combined with carboplatin in advanced non-small cell lung carcinoma: A phase II study. Proc Am Soc Clin Oncol 1996;15:(abstr 1200).

39   Hainesworth JD, Thompson DS, Urba WJ, Hon JK, Thompson KA, Hopkins LG, Grecco FA: One-hour paclitaxel plus carboplatin in advanced non-small cell lung cancer: Preliminary results of a multi-institutional phase II study. Proc Am Soc Clin Oncol 1996;15:(abstr 1131).

40   Evans WK, Stewart DJ, Tomiak E, et al: Carboplatin and paclitaxel by one-hour infusion for advanced non-small cell lung cancer. Proc Am Soc Clin Oncol 1995;14:1374(abstr 1156).

41   Sause WT, Scott C, Taylor S, et al: Radiation Therapy Oncology Group 88-08 and Eastern Cooperative Oncology Group 4588: Preliminary results of a phase III trial in regionally advanced, unresectable non-small cell lung cancer. J Natl Cancer Inst 1995;87:198–205.

42 Dillman RO, Seagren SL, Propert KJ, et al: A randomized trial of induction chemotherapy plus high-dose radiation versus radiation alone in stage III non-small cell lung cancer. N Engl J Med 1990;323:940–945.

43 LeChavalier T, Arriagada R, Quoix E, et al: Radiotherapy alone versus combined chemotherapy and radiotherapy in unresectable non-small cell lung carcinoma. Lung Cancer 1994;10:239–244.

44 Schaake-Koning C, Van den Bogaert W, Dalesio O, et al: Effects of concomitant cisplatin and radiotherapy on inoperable non-small cell lung cancer. N Engl J Med 1992;326:524–530.

45 Jeremic B, Shibamoto Y, Acimovic L: Randomized trial of hyperfractionated radiation therapy with or without concurrent chemotherapy for stage III non-small cell lung cancer. J Clin Oncol 1995;13:452–458.

46 Jeremic B, Shibamoto Y, Acimovic L, Milisarljenic S: Hyperfractionated radiation therapy with or without concurrent low-dose daily carboplatin-etoposide for stage III non-small cell lung cancer: A randomized study. J Clin Oncol 1996;14:1065–1070.

47 Choy H, Rodriguez FF, Koester S, et al: Investigation of taxol as a potential radiation sensitizer. Cancer 1993;71:3774–3778.

48 Liebmann J, Cook JA, Fisher J, Teague D, Mitchell JB: In vitro studies of taxol as a radiation sensitizer in human tumor cells. J Natl Cancer Inst 1994;86:441–446.

49 Choy H, Akerley W, Safran H, et al: Phase I trial of outpatient weekly paclitaxel and concurrent radiation therapy for advanced non-small cell lung cancer. J Clin Oncol 1994;12:2682–2686.

50 Choy H, Safran H: Preliminary analysis of a phase II study of weekly paclitaxel and concurrent radiation therapy for locally advanced non-small cell lung cancer. Semin Oncol 1995;22(suppl 9): 55–57.

51 Choy H, Akerley W, Safran H, Graziano S, Chung C, Cole B: Phase II trial of weekly paclitaxel and concurrent radiation therapy for locally advanced non-small cell lung cancer. Proc Am Soc Clin Oncol 1996;15(abstr 1098).

52 Langer C, Rosvold C, Kaplan R, et al: Induction therapy with paclitaxel (Taxol) and carboplatin (CBDCA) followed by concurrent chemotherapy in unresectable locally advanced non-small cell lung carcinoma. Preliminary report of FCCC 94-001. Proc Am Soc Clin Oncol 1996;15:376(abstr 1118).

53 Hudes R, Langer C, Movsas B, Schol J, Keenan E, Kilpatrick D, Young C, Curran W: Induction paclitaxel and carboplatin followed by concurrent chemoradiotherapy in unresectable locally advanced NSCLC: Report of FCCC 94-001. Proc Am Soc Clin Oncol 1997;16(abstr 1609).

54 Rosell R, Gomez-Codina J, Camps C, et al: A randomized trial comparing perioperative chemotherapy plus surgery with surgery alone in patients with non-small cell lung cancer. N Engl J Med 1994;330:153–158.

55 Roth J, Fossella F, Komaki R, et al: A randomized trial comparing perioperative chemotherapy and surgery with surgery alone in resectable stage IIIA non-small cell lung cancer. J Natl Cancer Inst 1994;86:673–680.

Corey J. Langer, MD, Fox Chase Cancer Center, Co-Director Thoracic Oncology, 7701 Burholme Avenue, Philadelphia, PA 19111 (USA)
Tel. (215) 728–2985, Fax (215) 728-3639, E-Mail cj.langer@oberon.fccc.edu

Schiller JH (ed): Updates in Advances in Lung Cancer. Prog Respir Res.
Basel, Karger, 1997, vol 29, pp 91–105

Chapter 6
••••••••••••••••••••••••

The Use of Gemcitabine in Non-Small-Cell Lung Cancer

Ulrich Gatzemeier[a], *Hans-Dieter Peters*[b]

[a] Department of Thoracic Oncology, Hospital Grosshansdorf/Hamburg, and
[b] Department of Immunopharmacology, Center of Pharmacology and Toxicology,
Medical University, Hannover, Germany

It is generally accepted that only 10% of patients with non-small-cell lung cancer (NSCLC) survive for more than 5 years, a survival rate that has not changed significantly in the last decade. Many NSCLC patients with advanced disease (stages IIIB and IV) are treated with cytostatic drugs to improve their overall prognosis and relieve symptoms. Currently, few cytostatics are effective for more than 15% of these patients: cisplatin, ifosfamide, mitomycin C, vinblastine, and vindesine [1]. Etoposide – with a low efficacy of around 10% – is often combined with cisplatin because it potentiates the activity of the latter. New cytostatics for NSCLC, with equivalent or better efficacy, are urgently needed to prolong survival with minimal toxicity.

Gemcitabine is a new nucleoside analog with activity against a range of solid tumor types. In animal models it has shown broad activity against tumors such as myeloma, breast carcinoma, ovarian carcinoma, lymphosarcoma, melanoma, and leukemia. In human xenograft models, gemcitabine was effective against colon carcinoma, head and neck tumors, breast cancer, non-small-cell and small-cell bronchial carcinoma, and pancreatic, stomach, and liver carcinomas.

Gemcitabine Monotherapy

The first series of phase I trials to study the toxicity profile of gemcitabine demonstrated that the therapeutic index after a once weekly dosing regime for 3 weeks, followed by a 1-week rest (a cycle of 28 days) was sufficient for phase II studies to be undertaken. This schedule, with a starting dose between 800 and 1250 mg/m^2 gemcitabine per week on days 1, 8 and 15, every 28 days, has been applied in various phase II studies throughout the world. The efficacy in gemcitabine-responsive patients (complete response, CR; partial response, PR) was validated by two independent Oncology Review Boards (ORBs; one in Europe, one in the USA); this means that 'response' was first defined by the investigators, but finally validated by the ORBs.

One phase II study was conducted in England and Denmark and published by Anderson et al. [2] in 1994. Gemcitabine, 800 mg/m^2 per week was administered as a 30-min infusion on days 1, 8 and 15 followed by a 1-week rest (one course). When it became clear that the incidence of undesirable side effects was low, from patient 54, the starting dose was raised to 1000 mg/m^2 per week. Of the 82 patients entered into this study, 71 (87%) were available for statistical evaluation; the remaining patients were not evaluable for various reasons. For example, 6 patients whose carcinomas progressed during the first treatment cycle did not receive the complete therapy. Other patients were excluded from the analysis because their lesions did not fulfill the criteria of the ORB. In other circumstances, some of these patients would have been classified as 'responders,' and so they brought down the efficacy rate in this study, which overall was 22.5% for the 71 assessable patients, or 19.5%, when considering all 82 patients originally entered with the 'intention to treat' (table 1).

The second gemcitabine phase II study that gave favorable results was undertaken at six centers in South Africa [3], with a design identical to that described above for the English/Danish collaboration, with a starting dose of 1000 mg/m^2 per week, later raised to 1250 mg/m^2 per week. A majority of patients had squamous epithelial carcinoma (table 2); 10 patients had large-cell lung cancer, a higher incidence than in other studies. Of the 84 patients entered into the study, 76 were eligible for evaluation. Of these, 15 (19.7%) showed an objective response, and 2 patients showed CR (table 1). Overall efficacy, taking into account all the patients entered in the study, was 17.9%. A response to gemcitabine was found among all histological NSCLC subtypes, including large-cell lung carcinoma.

Following the positive results achieved in these small studies, a larger multicenter phase II study to authorize the use of gemcitabine in NSCLC was initiated in Europe and Canada, to which 161 patients were recruited [4]. The study design was again identical to that described above, except that the starting dose was 1250 mg/m^2 per week. As in the English/Danish study, the majority of patients had adenocarcinoma (52.5%) or squamous epithelial cancer. No patients with large-cell lung cancer were entered. Of the 161 entered patients, 151 (95%) were eligible for statistical evaluation. Three patients achieved CR.

Table 1. Phase II studies with gemcitabine monotherapy

	Europe [2]	South Africa [3]	International [4]	USA [5]
Entered/evaluable	82/71	84/76	161/151	34/30
Response				
CR (%)	0	2 (3)	3 (2)	0
PR (%)	16 (22.5)	13 (17)	30 (20)	1
Overall response, %	22.5	17.9	21.9	3
95% CI, %	13.5–34	11.6–30.8	15.6–29.3	1-17
Duration of response, months				
Median	8.1	12.7	7.6	16.9[a]
Range	3.6–17.3[b]	4.6–14.8[b]	2.6–19.6	–
Median Survival, months	8.1	9.2	10.4	8.8

CI = Confidence interval.
[a] One patient.
[b] Censored data.

Table 2. Patient characteristics in phase II gemcitabine studies

	Europe [2] (n = 82)	South Africa [3] (n = 84)	International [4] (n = 161)	USA [5] (n = 34)
Gemcitabine dose, mg/m^2 per week × 3	800–1000	1000–1250	1250	800
Male/female	49/33	65/19	124/37	28/6
Age, years				
Median	57	59	59	64
Range	23–71	36–76	35–75	40–78
Performance status, %	0, 1, 2	0, 1, 2	0, 1, 2	0, 1
Histology				
Adenocarcinoma	53	22	70	18
Squamous	24	40	84	13
Other	5	22	7	3
Disease stage				
IIIA	18	15	7	
IIIB	24	34	50	26
IV	40	35	104	8

Table 3. Toxicity profile (% patients) in gemcitabine monotherapy phase II trials for NSCLC

	Europe (n=82) [2]		South Africa (n=84) [3]		International (n=161) [4]		USA (n=34) [5]	
	WHO 3	WHO 4	WHO 3	WHO 4	WHO 3	WHO 4	WHO 3	WHO 4
Anemia	4.9	0	7.2	0	4.4	0.6	3	6.1
Leukopenia	6.2	1.2	9.6	1.2	7	0	3	0
Neutropenia	17.5	5.9	25.3	3.6	19.6	5.7	11.5	3.8
Thrombocyto-penia	0	1.2	0	2.4	1.3	0	6.1	3
AST	3.3	1.6	3.6	0	6.3	2.5	9.1	0
ALT	10	7.5	1.2	2.4	10.8	1.9	12.5	3.1
Alkaline phosphatase	3.7	0	1.2	0	1.2	0	0	0
Bilirubin	0	0	0	0	0.6	0.6	0	0

AST = Aspartate aminotransferase; ALT = alanine aminotransferase.

Both histological subtypes demonstrated a response to gemcitabine. Altogether, 33 patients responded to the therapy: 21.9% (33/151) of the evaluable patients, 19.3% (33/161) of all entered patients (table 1).

The median duration of response in these three 'positive' gemcitabine studies varied from 7.6 to 12.7 months; median survival times (for both responders and nonresponders) were similar: 8.1–10.4 months.

In contrast to these three studies, an American phase II investigation gave a 'negative' result, in that only 1 of the 31 (3.2%) evaluable NSCLC patients responded to gemcitabine [5]. The starting dose of 800 mg/m^2 per week was lower than in two of the three above-mentioned studies. More important though, in this study there was a trend to dose reduction, in contrast to the dose increase reported in the other studies.

For all four studies, gemcitabine toxicity was low. Mild myelosuppression, with WHO grade 3 and 4 anemia, was recorded in 7.2 and 6.1%, respectively; where necessary, this could be corrected by conventional treatment with erythrocyte concentrates (table 3). WHO grade 3 and 4 neutropenia was observed in 25.3 and 5.7% of patients, respectively, but resulted in only a low incidence of infections. For the parameter infection, WHO grade 1 and 2 toxicity was documented in 7.9 and 1.7%, respectively; there was no grade 3 or 4 toxicity for this parameter. Thrombocytopenia (WHO grades 3 and 4) was rare (table 3). Moderate, asymptomatic hepatotoxicity (manifest or transient) was observed, but the elevation in liver enzymes was usually rapidly reversible. Other toxic effects, in general not dose limiting, included skin rashes, mild fever on the day of drug administration, proteinuria, lethargy, and peripheral edema. Flu-like symptoms were recorded for 23.3% of the patients; they were usually mild, of short duration, and rarely dose limiting. The cause has not been identified, but the symptoms could be relieved by the administration of paracetamol. For 1.9% of the patients, peripheral edema was rated the most serious undesirable side effect; there is no evidence for an association with cardiac, hepatic, or renal function disturbances. For those patients who received gemcitabine for up to 1 year, it is possible to examine the long-term toxicity profile of the drug, and there is no evidence of any significant cumulative toxicity.

Three further phase II gemcitabine studies for NSCLC have been undertaken in Japan [6, 7]. Sixty-eight patients were entered into the first study, of whom 19 had already received cytostatic treatment [6]. The other two studies [7] comprised only patients who had received no prior cytostatic therapy. The characteristics of the patients in the three Japanese studies are given in table 4.

The starting dose in the first study was 800 mg/m^2 per week on days 1, 8 and 15 followed by a 1-week rest (one course). A 20% dose increase to 1000 mg/m^2 per week was possible if there was no evidence of significant toxicity. In the two following studies, the starting dose was 1000 mg/m^2 per week, again with the possibility for elevation to 1250 mg/m^2 per week.

In the first study, none of the patients previously treated with cytostatics responded to gemcitabine, but 8 of the 49 evaluable chemonaive patients showed a PR (table 5). PRs were also recorded in the other two studies in chemonaive patients: 19 PR (26% response) and 14 PR (20.9% response) (table 5).

Relief of Clinical Symptoms with Gemcitabine

An important, and for the patient highly relevant, question is whether cancer-disease-related symptoms improve under cytostatic therapy. The protocols for the four non-Japanese phase II studies were therefore designed to document so-called secondary effects to describe clinically relevant signs of symptom alleviation. Certain practical criteria were defined to record an improvement in disease-related symptoms: (a) the duration of the symptom improvement had to be at least 4 weeks to be rated clinically relevant; (b) patients were classified (coded) according to the most severe symptom experience between clinical visits, independent of symptom duration, and (c) there was to be no simultaneous therapy for symptoms (e.g., glucocorticoids) during the study period (to the extent justifiable).

Table 4. Patient characteristics in the Japanese phase II studies with gemcitabine mono-therapy for NSCLC

	0201 [6]	0202 [7]	0203 [7]
Entered/evaluated	68/66	73/73	69/57
Chemonaive/pretreated	49/17	73/0	67/0
Gemcitabine starting dose, mg/m² per week × 3	800–1000	1000 (up to 1250 possible)	1000 (up to 1250 possible)
Male/female	50/16	54/19	50/17
Age, years			
Mean	66	68	66
Range	44–79	32–79	36–80
Performance status			
0, 1	42	59	40
2, 3	24	14	18
Cell type			
Adenocarcinoma	35	47	36
Squamous carcinoma	25	16	28
Others	6	10	3
Stage			
I, II	1	0	0
IIIA	25	12	12
IIIB	0	16	21
IV	40	45	34

In these four studies, 360 patients with advanced, inoperable NSCLC could be evaluated for the specific improvement of disease-related symptoms. Patients eligible for symptom documentation before and after the gemcitabine therapy were those demonstrating moderate to severe disease-related symptoms on entry into the studies. These patients also showed the clearest evidence of symptom improvement.

Pain relief was noted for 83% of the gemcitabine-responsive patients and in 32% of all patients (including those not responsive to gemcitabine) (table 6). The results for the symptom hemoptysis were particularly dramatic: improvement in 100% of responsive and 63% of all patients. In general, the data in table 6 demonstrate that for other symptoms typical in lung cancer, symptom improvement in gemcitabine-responsive patients was greater than in patients whose lesions showed no objective response. Nevertheless, even this latter group of patients demonstrated alleviation of disease-related symptoms.

In summary, the extent of symptom improvement in these gemcitabine phase II study patients was similar to the proportion showing an objective

Table 5. Results in the Japanese II NSCLC trials

	0201 [6]	0201 [6]	0202 [7]	0203 [7]
Prior chemotherapy	yes	no	no	no
Entered/evaluated	19/17	49/49	73/73	69/67
Response				
CR	0	0	0	0
PR	0	8	19	14
No change	7	31	35	34
Progression	8	5	15	15
Not evaluable	2	5	4	4
Overall response, %	0	16.3	26	20.9
95% CI, %	–	6.0–26.7	16.0–36.1	10.8–29.6
Duration of response	–	NA	NA	NA
Median survival	NA	NA	NA	NA

CI = Confidence interval; NA = not available.

Table 6. Gemcitabine effect on clinically relevant symptoms in NSCLC

Symptom	Responding patients			All eligible patients		
	patients with symptoms	improvement		patients with symptoms	improvement	
		n	%		n	%
Pain	18	15	83	129	41	32
Hemoptysis	8	8	100	40	25	63
Dyspnea	22	13	59	113	29	26
Coughing	29	23	79	140	62	44
Anorexia	6	4	67	35	10	29
Pleural effusion	7	4	57	30	10	33
Somnolence	9	4	44	39	16	41

improvement. These are significant results, given that NSCLC therapy for advanced stage IIIB and IV cancers is primarily for palliative purposes.

Combined Cytostatic Therapy with Gemcitabine – Phase II Studies

The phase II studies with gemcitabine monotherapy demonstrated that in NSCLC, the drug can achieve a reproducible efficacy of around 20% with an

Table 7. Combined gemcitabine and cisplatin phase II trials for NSCLC: patient characteristics

	South Africa [9]	Italy [10]	USA [11]	Canada [12]	UK/France [8]
Gemcitabine dose, mg/m²	1000 days 1, 8, 15	1000 days 1, 8, 15	1000 days 1, 8, 15	1000–2250 days 1, 8, 15	1000 days 1, 8, 15
Cisplatin dose, mg/m²	100 day 15	100 day 2	100 day 1	25–30 days 1, 8, 15	100 day 15
Male/female	26/14	41/7	17/9	33/17	50/10
Age, years					
Median	58	60	62	62	no data
Range	35–74	37–70	37–74	30–75	39–74
Performance status					
0	2		median		4
1	23		90		50
2	15	0			6
Cell type					
Adenocarcinoma	13	21	10	35	21
Squamous	21	22	10	8	29
Others	6	5	6	5	9
Stage					
IIIA	12	1	0	3	1
IIIB	14	21	5	12	40
IV	14	26	21	31	19

acceptable profile of undesirable side effects. The mechanism of gemcitabine action, described above, suggests that it might be worthwhile investigating what happens when it is combined with other cytostatic drugs causing DNA damage.

Gemcitabine with Cisplatin

Steward et al. [8] reported on an English/French phase I/II collaborative study investigating the safety and effectiveness of gemcitabine combined with cisplatin in previously untreated patients with NSCLC (table 7). Gemcitabine was given at a fixed dose of 1000 mg/m² per week on days 1, 8 and 15 every 28 days (one course); cisplatin was administered immediately after the third gemcitabine infusion (day 15) in each cycle. In the phase I part of the study with 16 patients (13 men/3 women, median age 58 years), the cisplatin dose was gradually increased from 60, through 75 to 100 mg/m² for each successive cohort of 3 patients, using an adaptive control algorithm. A further 50 patients were entered into the phase II study and administered 100 mg/m² cisplatin. Patients treated at this level of cisplatin in the phase I part of the study were included in the final analysis.

Table 8. Combined gemcitabine and cisplatin phase II trials for NSCLC: results

	South Africa [9]	Italy [10]	USA [11]	Canada [12]	UK/France [8]
Prior chemotherapy	none	none	none	none	none
Entered/evaluated	40/35	48/46	30/26	50/38	60/52
Efficacy					
CR	1	1	1	0	1
PR	15	25	10	12	21
Overall efficacy, %	46	54	42	31.6	42
Duration of response, months	no data	no data	no data	4	10.9
Median survival, months	no data	not yet evaluable	no data	not yet evaluable	11.3

Among the 66 recruited patients (phase I and II), there were 53 male patients; the age range was 39–74 years. Four patients had a performance status 0, 55 were status 1, and 7 were status 2. For the 60 evaluated patients, the NSCLC subtypes were as follows: adenocarcinoma 21, squamous epithelial carcinoma 29, large-cell cancer 9, and not specified 1. Tumor stages are shown in table 7. Fifty-five patients with 236 cisplatin infusions could be evaluated for the toxic effects of the combination, which were usually easy to control. The incidence of WHO grade 3/4 toxicity was as follows: neutropenia 52.5%, thrombocytopenia 27.1%, liver function enzymes (alanine aminotransferase/ aspartate aminotransferase) 1.8%, alopecia 1.5%, fever 3.1%, nausea and vomiting 46.1%. No severe side effects in the urinary tract were recorded. The therapeutic efficacy in the phase II part of the trial was calculated for 52 of the 60 entered patients. There was 1 CR and 21 PRs, giving an efficacy rate of 42.0% (95% confidence interval, CI: 28.7–56.8%; table 8). The authors concluded that the gemcitabine/cisplatin combination was well tolerated and demonstrated a clear cytostatic effect. There was no demonstrable increase in toxicity and further optimization of the combination was recommended. Use of this combination requires only one night of hospitalization every 4 weeks.

Abratt et al. [9] analyzed the toxicity and efficacy of the gemcitabine/ cisplatin combination in a phase II trial of patients with local advanced or metastasized NSCLC (table 7). Again, gemcitabine was administered on days 1, 8 and 15 (followed by a 1-week rest) at 1000 mg/m^2 per week; the cisplatin dose was 100 mg/m^2 administered on day 15. Only previously untreated patients were entered in the study, whose carcinoma lesion was measurable on computed tomography, and whose creatinine clearance reached at least 60 ml/min. Forty patients were entered of whom 35 were evaluable, having received at least two courses of therapy. Their average age was 58 years (range 35–75). Their cancer staging was as follows: IIIA 21%, IIIB 39%, and IV 39%. One patient demonstrated a CR; a PR was recorded for 15 patients (46%; table 8).

Among the patients ascertained for toxicity, the following WHO grade 3 and 4 toxicities were recorded: neutropenia (29 and 18%), thrombocytopenia (29 and 6%), and nausea (53 and 0%). Serum creatinine rose in 12% of the patients (WHO grade 1).

Crino et al. [10] reported a multicenter (12 institutions) phase II Italian study of the gemcitabine/cisplatin combination. Of the 48 previously untreated patients entered into the study (see table 7), 46 were evaluable. Twenty-one patients had a locally advanced, unresectable stage IIIB lesion and 26 patients had disseminated stage IV cancer. Gemcitabine was administered at a fixed dose of 1000 mg/m^2 per week (days 1, 8 and 15) with a 1-week rest (one course). On the basis of experimental and clinical evidence for synergism between the two substances, cisplatin (100 mg/m^2) was administered on the second day of each cycle, shortly after the first gemcitabine dose.

Safety and efficacy were evaluated for those patients with measurable lesions who had received at least two courses of the combination therapy. Among the 46 evaluated patients there was 1 CR and 25 PRs (54%; 95% CI: 44–72%). According to staging, these remissions were as follows: stage IIIB – 11 (52%; 95% CI: 31–73%); stage IV – 1 CR and 14 PR (57%; 95% CI: 41–77%). Following resection and radiation therapy, 2 patients with a PR are currently disease free. This study is still running, and so the duration of response and survival have not yet been established.

Thrombocytopenia was the most serious side effect, 51% of patients demonstrating WHO grade 3/4 toxicity. It was usually of short duration, but nevertheless necessitated the interruption of gemcitabine administration on day 15 in 90 courses. No serious bleeding was recorded. In general, nonhematological toxicity was mild; there was one acute but reversible kidney failure.

A US phase II study again tested the combination of gemcitabine (1000 mg/m^2 per week on days 1, 8 and 15) and cisplatin (100 mg/m^2, day 1) [11] (table 7). Entered into the study were 30 previously untreated patients with advanced NSCLC, of whom 26 (17 men, 9 women; median age 62 years, range 37–74) were evaluated for efficacy and toxicity. The other 4 patients still remain to be evaluated. Eleven patients responded to the therapy (1 CR, 10 PR) – an overall response rate of 42% (table 8).

A Canadian phase I/II study [12] investigated the combination of gemcitabine (1000–2250 mg/m^2 per week, days 1, 8 and 15) and cisplatin (25–30 mg/m^2, days 1, 8 and 15) (table 7) in 50 previously untreated patients with advanced NSCLC. At the time the report appeared, 38 patients were evaluable for efficacy and tolerance. Thirty-one patients had stage IV disease with pronounced metastasis. Twelve patients achieved a PR – the overall response rate was 31.6% (95% CI: 17–49%). The mean duration of response was longer than 5 weeks and the time to therapy failure, 15 weeks. A response to the combined gemcitabine/cisplatin therapy was seen at all dose levels, meaning that, at least with a small patient collective, a definable dose-response relationship for gemcitabine is not demonstrable.

Gemcitabine plus Carboplatin

Carmichael et al. [13] undertook a study to define the maximum tolerated dose (MTD) of carboplatin in the combination of gemcitabine and carboplatin for previously untreated NSCLC patients with locally advanced or metastatic disease. Gemcitabine was administered as a fixed dose of 1000 mg/m^2 per week on days 1, 8 and 15 followed by a 1-week rest (one course). Carboplatin was given immediately before the gemcitabine infusion on day 1 of each cycle (a short-duration infusion of 30 min). The dose of carboplatin to be used was calculated from the area under the curve (AUC) using creatinine clearance and the formula of Calvert. Doses were escalated according to an adaptive control algorithm. The MTD was determined in 9 patients as the point where significant toxicity greater than 30% was demonstrated, i.e., at the appearance of nonhematological WHO grade 3 or hematological WHO grade 4 toxicity or manifestation of severe nausea or vomiting.

To date, 13 patients have been entered in the study (10 men, 3 women, age range 41–72 years), the majority (11 patients) with performance status 1. Histologically, the NSCLC subtypes were as follows: adenocarcinoma (n = 3), squamous cell carcinoma (n = 6), and large-cell cancer (n = 3). One patient was stage IIIA, 5 patients stage IIIB and 7 patients were at stage IV.

Three patients were taken into the study at the initial carboplatin dose level (AUC 4.0 mg/ml·min) and 10 at the AUC level of 5.2 mg/ml·min. No significant toxicity was registered at the lower dose, but WHO grade 3 and 4 toxicity appeared at the higher dose: leukocytes (4 and 1 patient, respectively), neutrophils (7 and 2 patients, respectively), and thrombocytes (1 and 2 patients, respectively). The provisional results indicate that carboplatin exerts no direct effect on the pharmacokinetics of gemcitabine or its chief metabolite dFdU. Significant activity was recorded at the AUC level of 5.2 mg/ml·min, but this interaction appears to be pharmacodynamic rather than pharmacokinetic. A future study will investigate the effect of administering the drugs in the reverse order.

Gemcitabine plus Ifosfamide

Eberhard et al. [14] and Wilke et al. [15] are currently conducting a phase I/II dose-finding study with the gemcitabine/ifosfamide combination in previously untreated patients with advanced NSCLC. These drugs were se-

Table 9. Combined gemcitabine and ifosfamide phase II trial: patient characteristics

Gemcitabine dose, mg/m^2	1000 on days 1, 8 and 15
Ifosfamide dose, mg/m^2	1500 on days 8–12
Treated/evaluated	56/50
Median performance status	80
Cell histology	
Adenocarcinoma	10
Squamous cell	36
Others	10
Stage	
IIIA	1
IIIB	8
IV	47

lected as combination partners because they demonstrate different mechanisms of activity and toxicity profiles. Gemcitabine is adminstered in a fixed dose of 1000 mg/m^2 per week (days 1, 8 and 15) with a 1-week rest (one course); ifosfamide is infused on days 8–12 of each cycle. In the phase I part of the study, ifosfamide was administered at escalating doses of 1200 (dose level 1), 1500 (dose level 2), and 1800 (dose level 3) mg/m^2 to successive patient cohorts (a minimum of 3 patients per cohort with no individual dose escalation).

Eighteen patients were entered into the phase I study. Of 11 currently evaluable patients, 6 have registered a PR (54.5%). At the ifosfamide dose level 1, of 5 evaluable patients, 2 demonstrated a PR. Two patients had to break off the study at this level, 1 demonstrating exanthema (WHO grade 2) and 1 after the second gemcitabine infusion because of an ifosfamide-related WHO grade 3 neurotoxicity. At the ifosfamide dose level 2, 4 of the 5 patients were evaluable (2 PR); 1 patient was disqualified due to a protocol violation (condition following radiotherapy at study entry). Six patients were evaluable at the ifosfamide dose level 3 – 2 PRs were documented. The remaining patients remain to be evaluated. At this level, leukopenia was dose limiting. The investigators [14, 15] conclude that the ifosfamide/gemcitabine combination is clearly active in NSCLC patients. The study demonstrated a MTD of 1500 mg/m^2 for ifosfamide administered simultaneously with gemcitabine, which was usually well tolerated.

Dose level 2 (ifosfamide: 1500 mg/m^2 on days 8–12; gemcitabine: 1000 mg/m^2 over 3 weeks) was selected for the phase II study for which 56 untreated NSCLC patients in three centers (Essen, Grosshansdorf, and Heidelberg) have been entered. So far, 50 of these patients can be evaluated (table 9). In the

Table 10. Combined gemcitabine and ifosfamide phase
II trial: results

Remission rate	
CR/PR	11/50 (22%)
No change	32/50 (64%)
Duration of remission, months	5.1
Toxicity, %	
Neutropenia	
WHO 3	32
WHO 4	15
Thrombopenia	
WHO 3	18
WHO 4	3
Nausea/vomiting	
WHO 3	15
WHO 4	0

'intent-to-treat' analysis, 11 of 50 (22%) of the evaluable patients demonstrated a PR, in 32 (64%) the disease was stable. Seven patients showed progression (table 10). The dominant toxic effect was a readily reversible myelosuppression; other side effects were in general mild to moderate.

Phase III Study: Gemcitabine versus Cisplatin/Etoposide

The first results of a prospective, randomized phase III study conducted by a German group [16] were presented at the EMSO Congress in Vienna in November 1996 (table 11). Gemcitabine monotherapy is being compared to combination treatment with cisplatin and etoposide. The gemcitabine dose, at 1000 mg/m^2 was somewhat lower than in most phase II studies. The cisplatin/etoposide dosing was standard. For the 107 currently evaluated patients, similar activity, calculated for various prognostic factors (age, sex, stage), in terms of remission rate and progression-free survival was seen for the gemcitabine monotherapy and the combined cisplatin/etoposide treatment. However, the gemcitabine monotherapy was clearly tolerated much better, demonstrating a clear advantage for gemcitabine therapy in terms of toxicity and quality of life.

In summary: The preclinically demonstrated activity of the new pyrimidine antimetabolite, gemcitabine, was confirmed in all but one phase II study with NSCLC patients. Gemcitabine achieved similar response rates to the four substances – cisplatin, ifosfamide, mitomycin C, and vindesine – currently

Table 11. Phase III study: gemcitabine versus cisplatin/etoposide

Therapy	Patients	Stage III/IV	Remission[a], %	Time to progression, months
Cisplatin 100 mg/m^2, day 1	55	12/43	6.9 (1.9–19.7)	3.7
Etoposide 100 mg/m^2, days 1–3				
Gemcitabine 1000 mg/m^2, days 1, 8, 15	52	13/39	17.2 (8.6–29.4)	4.2

[a] Range in parentheses: 95% CI.

prescribed for NSCLC. Gemcitabine, with its novel mode of action, extends the range of cytostatic substances effective for NSCLC, and demonstrates, moreover, an acceptable profile of undesirable side effects.

Attention should be drawn to the fact that the results in the NSCLC trials described above have been validated by independent ORBs; in other words, the 'response' was defined not only by the clinical investigators, but also by an independent body of experts.

It remains to be ascertained whether the response of NSCLC patients to gemcitabine is dose dependent. A meta-analysis of the responses and dose intensities in the first phase II studies from Europe, South Africa, the USA, and an international study, is underway. The results indeed suggest that dose is significant. Patients receiving less than 900 mg/m^2 per week of gemcitabine showed no response. Between 900 and 1099 mg/m^2 per week, the response rate was 16.7%; between 1100 and 1299 mg/m^2 per week, this rose to 26.9%. On the other hand, further dose escalation to more than 1300 mg/m^2 per week produced a response rate of just 18.9%. As yet, these differences are not statistically significant, probably because of the small patient numbers in each subgroup.

In the first phase III study comparing gemcitabine monotherapy with cisplatin/etoposide combination therapy, they demonstrated similar efficacy but there was a clear advantage of the monotherapy in terms of toxicity and quality of life.

References

1 Ginsberg RJ, Kris MG, Armstrong JG: Cancer of the lung; in DeVita VT, Hellman S, Rosenberg SA (eds): Cancer: Principles and Practice of Oncology, ed 4. Philadelphia, Lippincott, 1993, pp 673–723.

2 Anderson H, Lund B, Bach F, Thatcher N, Walling J, Hansen HH: Single-agent activity of weekly gemcitabine in advanced non-small cell lung cancer: A phase II study. J Clin Oncol 1994;12: 1821–1826.

3 Abratt RP, Bezwoda WR, Falkson G, Goedhals L, Hacking D, Rugg TA: Efficacy and safety profile of gemcitabine in non-small cell lung cancer: A phase II study. J Clin Oncol 1994;12:1535–1540.

4 Gatzemeier U, Shepherd F, LeChevalier T, Weynants P, Cottier B, Groen HJM, Rosso R, Mattson K, Cortes-Funes H, Tonato M, Burkes RL, Gottfried M, Voi M: Activity of gemcitabine in patients with non-small cell lung cancer: A multicentre, extended phase II study. Eur J Cancer 1996;32A: 243–248.

5 Kaye SB: Gemcitabine: Current status of phase I and II trials. J Clin Oncol 1994;12:1527–1531.

6 Fukuoka M, Kurita Y, Niitani H and the Japanese Gemcitabine Lung Cancer Study Group: Minisymposium at the 7th World Conference on Lung Cancer, June–July 1994, Colorado Springs, Colo, USA.

7 Nakai Y, Takada M, Yokoyama A, Negore S, Kurita Y, Fukuoka M, Niitani H: Results of phase II studies of gemcitabine in patients with non-small cell lung cancer in Japan. J IASLC 1994; 120(abstr 460).

8 Steward WP, Dunlop DJ, Cameron C, Talbot DC, Kleisbauer J-P, Thomas P, Guerin JC, Perol M, Samson G, Dabouis G, Lacroix H: Phase I/II study of cisplatin in combination with gemcitabine in non-small cell lung cancer. Proc Am Soc Clin Oncol 1995;14:357(abstr 1064).

9 Abratt RP, Bezwoda WR, Goedhals L, Hacking DJ: A phase 2 study of gemcitabine with cisplatin in patients with non-small cell lung cancer. Proc Am Soc Clin Oncol 1995;14:375(abstr 1159).

10 Crinò L, Scagliotti G, Marangolo M, Figoli F, Clerici M, De Marinis F, Salvati F, Cruciani G, Dogliotti L, Cocconi G, Paccagnella A, Adamo V, Incoronato P, Scarcella L, Mosconi AM, Tonato M: Cisplatin-gemcitabine combination in non-small cell lung cancer: A phase II study. Proc Am Soc Clin Oncol 1995;14:352(abstr 1066).

11 Sandler AB, Ansari R, McClean J, Fisher W, Dorr A, Einhorn LH: A Hoosier Oncology Group phase II study of gemcitabine plus cisplatin in non-small cell lung cancer. Proc Am Soc Clin Oncol 1995;14:357.(abstr 1089)

12 Shepherd FA, Crump M, Burkes R, Cormier Y, Feld R, Strack T: Gemcitabine and weekly cisplatin in NSCLC. Lilly Oncology Global Medical Conference, May 1995, Indianapolis, Ind, USA.

13 Carmichael J, Allerheiligen S, Walling J: A phase I/II study of gemcitabine and carboplatin in NSCLC. Proc Am Soc Clin Oncol 1995;14:351(abstr 1065).

14 Eberhard W, Wilke H, Manegold CH, Gatzemeier U, Blatter J, Drings P, Seeber S: Phase I dose finding study of gemcitabine and ifosfamide in advanced non-small cell lung cancer. Proc Am Soc Clin Oncol 1995;14:351(abstr 1063).

15 Wilke HJ, Manegold C, Gatzemeier U, Eberhard W, Blatter J, Drings P, Seeber S: Gemcitabine and ifosfamide in advanced NSCLC: A phase I-II study. Lilly Oncology Global Medical Conference, May 1995, Indianapolis, Ind, USA.

16 Manegold C, von Pawel J, Conte PF, Dornhoff W, van Walree NC, ten Bokkel Huinink WW, Drings P: Randomised phase II study of gemcitabine versus cisplatin/etoposide in patients with advanced non-small cell lung cancer. Ann Oncol 1996;7(suppl 5):3(abstr 30).

Ulrich Gatzemeier, MD, Department of Thoracic Oncology, Hospital Grosshansdorf/Hamburg, Center of Pneumology and Thoracic Surgery, Woehrendamm 80, D–22927 Grosshansdorf/Hamburg, (Germany)

Schiller JH (ed): Updates in Advances in Lung Cancer. Prog Respir Res.
Basel, Karger, 1997, vol 29, pp 106–116

Chapter 7
•••••••••••••••••••••••••

The Use of Docetaxel in Non-Small Cell Lung Cancer

James F. Bishop, Stephen J. Clarke

Sydney Cancer Centre, Royal Prince Alfred Hospital, Sydney, N.S.W., Australia

Introduction

Paclitaxel and docetaxel are two members of a new class of anticancer drugs known as taxoids which interact with the microtubular structure of the mitotic spindle. Taxoids promote the rate and extent of tubulin assembly into stable microtubules thus preventing tubulin depolymerization, leading to interruption of cell division at G_2/M and apoptotic cell death [1, 2]. Both paclitaxel and docetaxel share the same microtubule binding site. In comparison to paclitaxel, docetaxel is twice as potent as an inhibitor of microtubule depolymerization and has a 1.9-fold greater affinity for the microtubule binding site [3, 4]. Comparative cellular efflux studies of the taxoids demonstrate that docetaxel is retained intracellularly, with a 3 times longer half-life than paclitaxel [5]. It is not known whether such pharmacological differences are clinically relevant when comparing the relative toxicity and efficacy of paclitaxel and docetaxel.

Docetaxel has potent in vitro activity against a wide range of murine tumors and human tumor xenografts in mice [1, 6–8]. Activity was demonstrated against human ovarian (OVCAR-3), lung (LX-1), breast (Calc 18 and MX-1) and colon (KM20L2) carcinoma xenografts, and SKMEL-2 melanoma. Docetaxel was also active against cisplatin-insensitive OvPe ovarian cancer [9]. Docetaxel-resistant cell lines are also resistant to vinblastine, but show minimal to moderate levels of cross-resistance to vincristine, etoposide, cisplatin or doxorubicin and no cross-resistance with camptothecin or 5-fluorouracil [10, 11]. Docetaxel can inhibit replication of paclitaxel-resistant murine macrophages [2]. In vitro synergism was seen with the following combinations; docetaxel-vinorelbine, docetaxel-cyclophosphamide, docetaxel-5-

Table 1. Phase I studies of docetaxel

Group (first author)	Infusion time h	Schedule	Patients n	MTD mg/m^2	Dose-limiting toxicities
Extra [13]	1–2	Q3W	65	115	Neutropenia
Burris [14]	2	Q3W	18	115	Neutropenia Skin reaction
	6	Q3W	40	100	Neutropenia Mucositis
Bissett [15]	24	Q3W	30	90	Neutropenia Mucositis
Pazdur [16]	1	Days 1–5 Q3W	39	16×5 (or 80)	Neutropenia Mucositis
Tomiak [17]	1	Days 1+8	33	55×2 (or 110)	Neutropenia
Aapro [18]	1	Q3W	10	N/A	Neutropenia

fluorouracil, and docetaxel-vincristine [1]. The least overlap in toxicity was with docetaxel-vincristine.

Phase I Trials

For phase I trials, docetaxel was initially formulated at a concentration of 15 mg/ml in a diluent of 50% polysorbate (Tween) 80 and 50% ethanol. Subsequently, docetaxel has been re-formulated as a 40 mg/ml solution in polysorbate 80 [12].

Five phase I dose-escalation studies and one dose-confirmation study involving 256 patients have been conducted in North America and Europe [12–18] (table 1). The principal dose-limiting toxicity was neutropenia. NCI grade 4 neutropenia occurred in 52% of patients in the dose range 70–94 mg/m^2 and in 80% in the dose range 95–105 mg/m^2. Neutropenia was uncommon in doses not less than 70 mg/m^2. Febrile neutropenia occurred in approximately 10% of patients [19]. Mucositis was seen in 10% of 1 to 2-hour docetaxel infusions and correlated with neutrophil nadir [19]. The incidence of mucositis varied with the schedule of administration of docetaxel. Using docetaxel daily $\times 5$, 46% of patients developed mucositis [16] compared with 28 and 44% after 6- and 24-hour infusions, respectively [14, 15].

Other toxicities encountered included fluid retention, skin reactions, most commonly pruritus, hypersensitivity, neuropathy, diarrhea and asthenia. The fluid retention mainly affected the face and hands, but sometimes was associated with weight gain and pleural effusions [17].

Attempted amelioration of these toxicities occurred as part of the phase II evaluation. There was substantial improvement in skin reactions, hypersensitivity and fluid retention after pre- and posttreatment administration of corticosteroids for 3–5 days [20, 21].

The major route of elimination is by fecal excretion via the biliary route [22, 23]. Docetaxel is extensively metabolized by the CYP3A family of cytochrome P_{450} liver enzymes to four major metabolites which are 30- to 140-fold less effective, and less toxic, than the parent compound [24, 25]. It is becoming increasingly clear that toxicity is substantially enhanced when full-dose docetaxel is given in the presence of liver impairment [26, 27]. Thus, dose reductions are required for starting doses in patients with moderate elevation of liver enzymes and no dose can safely be recommended in patients with an elevated bilirubin. Up to 9-fold variations in biotransformation rates of docetaxel have been demonstrated in vitro using a human liver microsome library [28]. Thus, intrinsic inter-patient variation in microsome enzyme levels may contribute to variable toxicity after administration of docetaxel, even in the absence of hepatic metastases. There is also the potential for pharmacokinetic interaction of docetaxel with other commonly used drugs, such as erythromycin, ketoconazole, nifedipine and midazolam which also interact with CYP3A.

Objective responses were seen in 4 patients with non-small cell lung cancer (NSCLC) in the phase I studies conducted by Tomiak et al. [17], Burris et al. [14] and Extra et al. [13]. Overall, phase I studies suggested that prolonged infusions or daily $\times 5$ dosing did not permit greater dose intensity and were more toxic with mucositis becoming dose limiting. Thus, the 1-hour infusion schedule was taken to phase II studies in NSCLC.

Phase II Single-Agent Studies of Docetaxel in NSCLC

Phase II trials of docetaxel have used 100 mg/m², as a 1-hour intravenous infusion, every 3 weeks in three studies [29–31], and at 60 and 75 mg/m², in two others [32, 33]. Overall, the objective response was 28%, but was 31% in studies where patients received docetaxel at a recommended dose of 100 mg/m² [29–33] (table 2). The median duration of response varied considerably from 3.3 to 9.1 months. These results compare quite favourably to responses obtained in phase II trials of paclitaxel, CPT-11, gemcita-

Table 2. Docetaxel as a single agent in untreated NSCLC

Group (first author)	Evaluable patients n	Objective response %	Median duration of response months	Median survival months
Fossella [29]	39	33	3.3	11
Cerney [30]	35	23	9.0	11
Francis [31]	29	38	5.3	6.3
Miller [32][a]	20	25	9.1	–
Kudo [33][b]	72	25	3.3	–
Total	195	28		

[a] Miller et al. [32] used docetaxel at 75 mg/m^2.
[b] Kudo et al. [33] used docetaxel at 60 mg/m^2.

Table 3. Single-agent activity of newer agents in NSCLC

Drug	Patients, n	Objective response %	References
Docetaxel	195	28	29–33
Paclitaxel	49	22	34, 35
CPT-11	150	30	36, 38
Gemcitabine	306	21	39–41
Vinorelbine	207	31	42–44

bine or vinorelbine in previously untreated patients with NSCLC [29–44] (table 3).

Fossella et al. [45] reported that 9 of 42 (21%) patients with platinum-resistant NSCLC responded to second-line docetaxel. A confirmatory study of single-agent docetaxel in France reported a 27% response in patients who were refractory or had relapsed following platinum therapy for NSCLC [46]. The duration of response was approximately 6 months. These results are particularly exciting since previous ECOG studies have shown extremely poor responses to second-line MVP therapy after failure of single-agent platinum in NSCLC [46]. Furthermore, paclitaxel and CPT-11 both appear inactive after platinum failure [37, 47]. These data may be indicative of important activity for docetaxel in NSCLC, and suggest that it may differ from paclitaxel for this tumor type.

Cisplatin/Docetaxel Combinations in NSCLC

The activity of docetaxel in cisplatin-resistant cell lines [1] and its clinical activity in cisplatin-resistant NSCLC patients [45] provided a rationale for the investigation of the combination, cisplatin/docetaxel. However, preclinical data, such as the combination toxicity index, suggested that the two drugs in this combination may have substantial overlapping toxicity [1].

A phase I dose-escalating study was begun in Australia with docetaxel given over 1 h immediately followed by cisplatin over 1 h, with postinfusion hydration [49]. Doses of docetaxel/cisplatin of 50/75, 75/75, 75/100 and 100/75 respectively were used.

Of 24 previously untreated patients entered, all were evaluable for toxicity and 18 for response. The maximum tolerated dose schedules were docetaxel 75 mg/m^2 with cisplatin 100 mg/m^2 with dose-limiting toxicities in 5 of 6 patients which were both febrile neutropenia and nonhaematological toxicities, principally diarrhea. Two of 24 patients developed neutropenic enterocolitis. Dose-limiting toxicity, mainly neutropenia, was also seen with docetaxel 100 mg/m^2 and cisplatin 75 mg/m^2. Pharmacokinetics of both drugs were consistent with results from single-agent studies, suggesting no major pharmacokinetic interaction between the two drugs. Neutropenia was related to docetaxel area under the plasma concentration versus time curve (AUC). An alternative schedule was investigated, with cisplatin being administered over 3 h commencing 3 h after docetaxel, but the toxicity was similar. Independently reviewed responses occurred in 8/18 (44%) of patients, most following 75 mg/m^2 of both drugs. This study recommended a phase II dose of docetaxel 75 mg/m^2 and cisplatin 75 mg/m^2.

A Dutch phase I study in miscellaneous tumors reported 91% grade 3 and 4 neutropenia with the same combination, but with doses pushed to docetaxel 85–100 mg/m^2 and cisplatin 75 mg/m^2 [50].

Phase II studies in NSCLC with this combination have been performed in Australia, USA and France [51–53]. In the Australian study, NSCLC patients, previously untreated with chemotherapy, ECOG performance status 0–2 and normal liver, bone marrow and renal function were eligible [52]. Patients received docetaxel 75 mg/m^2 i.v. over 1 h immediately followed by cisplatin 75 mg/m^2 i.v. over 1 h and postchemotherapy hydration. All patients received premedication including dexamethasone 20 mg orally, 12 and 6 h prior to chemotherapy and ondansetron and promethazine immediately prior to treatment. A total of 47 patients were entered on this study. The objective response rate was 30% for all patients, and 39% for patients evaluable for response. The median duration of objective response was 5.3 months and time to progression for all patients was 4 months. The median survival was 9.2 months. The Australian and French phase II studies have recently been compared [51]. In spite of the higher cisplatin dose used in France, the objective responses are remarkably similar with 30% of all French patients on study responding or 34% of those evaluable for response. However, the duration of response appeared longer in the French study at 11 months. Authors speculated that the longer duration of response may have reflected the use of a higher dose of cisplatin in the French study.

The combination was reasonably well tolerated with the phase I toxicity confirmed in the two phase II studies. The major toxicity was neutropenia, 13–15% febrile neutropenic episodes, 6–13% severe diarrhea and 0–4% severe mucositis seen in both studies. However, no patient developed neutropenic enterocolitis.

The combination docetaxel and cisplatin has been piloted by ECOG and is included as one of its experimental arms in its current phase III trial. This trial is for good performance status (ECOG 0–1), previously untreated patients with locally advanced or metastatic (stage IIIb or IV) NSCLC. The study will randomize patients to one of four combinations, cisplatin/paclitaxel, cisplatin/docetaxel, carboplatin/paclitaxel or cisplatin/gemcitabine.

This study builds on a recently reported ECOG study which showed that cisplatin/paclitaxel was superior to cisplatin/etoposide with a 1-year survival rate of 37% for the paclitaxel combination [54]. In this trial, the objective response rates were 27 and 32% for the two paclitaxel/cisplatin arms compared with only 12% for cisplatin/etoposide. On both paclitaxel arms, the median survival was approximately 10 months. While these gains appear modest, they probably represent a real advance in the treatment of lung cancer. The results of this trial may not be confirmed when the final results are available of a similar EORTC trial comparing cisplatin/teniposide with cisplatin/paclitaxel [55].

The proposed ECOG trial will allow the comparison of a number of active new drugs combined with a platinum. It will also be one of the first direct randomized comparisons of paclitaxel and docetaxel in combination with cisplatin. It will lead to a better understanding of the relative efficacy, toxicity and long-term results of these four promising combinations.

Docetaxel with Nonplatinum Drugs in NSCLC

The relatively new drugs with promising activity in NSCLC listed in table 3, are currently being tested in cisplatin combinations. These trials will require large numbers of patients and some years before the results are fully known. A further area of great potential is to investigate these new agents together in nonplatinum combinations. The rationale for this approach is based on the relative inactivity of platinum in NSCLC when compared to the proven activity of these new drugs.

The vinca alkaloids, vinblastine and vindesine have been in use as relatively weak agents in lung cancer for many years. A new synthetic approach has resulted in a new vinca alkaloid, vinorelbine, synthesized with an eight-member catharanthine ring and possessing impressive preclinical anticancer activity [56]. As a single agent in phase II trials, vinorelbine possesses impressive activity in previously untreated patients with NSCLC with 31% of patients achieving an objective response [42–44]. However, in large randomized trials, the response rate as a single agent is between 12 and 16% [57, 58]. These randomized trials reported superiority of the combination cisplatin/vinorelbine compared to vinorelbine alone, cisplatin alone or cisplatin/vindesine [56–59]. This combination, vinorelbine/docetaxel, is currently being studied [60, 61]. Kourousis et al. [60] reported phase II results with a 41% response rate in 37 NSCLC patients treated with vinorelbine/docetaxel and G-CSF. A novel approach of docetaxel alternating with cisplatin/vinorelbine produced similar preliminary results with a 44% objective response [62]. Of particular interest in this study is the moderate toxicity reported.

Future Directions for Docetaxel in NSCLC

The combinations suggested above of docetaxel combinations with other drugs such as CTP-11, topotecan, gemcitabine, etoposide or ifosfamide are yet to be reported. However, they hold great hopes that more optimal combinations will be discovered.

An exciting prospect for docetaxel would be to study it in early stage disease. Although it remains controversial, randomized studies by LeChevelier et al. [64], Dillman et al. [63] and the Radiotherapy Oncology Group [65] all show that induction with cisplatin-based chemotherapy improves survival compared to radiotherapy alone in locally advanced, stage IIIb, NSCLC. Docetaxel alone or in combination appears to be more active than the older drugs or combinations used in these trials. The use of docetaxel as a single agent or in combination in randomized studies of early stage disease now appears warranted.

Concurrent chemoradiotherapy may produce improved local control and potentially increased survival in locally advanced NSCLC [65, 66]. The taxanes are potent radiation sensitizers enhancing radiation-induced cytotoxicity synergistically [67, 68]. Docetaxel and radiation have been successfully combined in a phase I study using weekly docetaxel with radiotherapy 60 Gy to the chest [68]. This approach now needs careful clinical evaluation in locally advanced NSCLC but holds considerable promise for improving local control.

References

1   Bissery MC: Preclinical pharmacology of docetaxel. Eur J Cancer 1995;31A(suppl 4):1–6.
2   Ringel I, Horwitz SB: Studies with RP 56976 (Taxotere): A semi-synthetic analog of taxol. J Natl Cancer Inst 1991;83:288–291.
3   Gueritte-Voegelein F, Guenard D, Lavelle F, Le Goff MT, Mangatal L, Potier P: Relationships between the structure of taxol analogues and their antimitotic activity. J Med Chem 1991;34: 992–998.
4   Diaz JF, Andreu JM: Assembly of purified GDP-tubulin into microtubules induced by RP 56976 and paclitaxel: Reversibility, ligand stoichiometry and competition. Biochemistry 1993;32:2747–2755.
5   Riou JF, Petitgenet O, Combeau C, Lavelle F: Cellular uptake and efflux of docetaxel (Taxotere) and Paclitaxel (Taxol) in P388 cell line (abstract). Proc Am Assoc Cancer Res 1994;35:385.
6   Bissery MC, Guenard D, Gueritte-Coegelein F, Lavelle F: Experimental antitumour activity of Taxotere (RP 56976, NSC 628503), a taxol analogue. Cancer Res 1991;51:4845–4852.
7   Dykes DJ, Bissery MC, Harrison SD, Waud WR: Response of human tumour xenografts in athymic nude mice to docetaxel (RP 56976, Taxotere). Invest New Drugs 1995;13:1–11.
8   Nicoletti MI, Lucchini V, D'Incalci M, Giavazzi R: Comparison of paclitaxel and docetaxel activity on human ovarian carcinoma xenografts. Eur J Cancer 1994;30A:691–696.
9   Boven E, Venema O, Gaberscek E, Erkelens CAM, Bissery MC, Pinedo HM: Antitumour activity to Taxotere (RP 56976, NSC 628503), a new Taxol analog, in experimental ovarian cancer. Ann Oncol 1993;4:321–324.
10 Riou JF, Petitgenet O, Aynie I, Lavelle F: Establishment and characterization of docetaxel (Taxotere) resistant human breast carcinoma (Calc 18/TXT) and murine leukemic (P388/TXT) cell lines (abstract). Proc Am Assoc Cancer Res 1994;35:339.
11 Hill BT, Whelan RDH, Shellard SA, McClean S, Hosking LK: Differential cytotoxic effects of docetaxel in a range of mammalian tumor cell lines and certain drug-resistant sublines in vitro. Invest New Drugs 1994;12:169–182.
12 Aapro M, Bruno R: Early clinical studies with docetaxel. Eur J Cancer 1995;31A(suppl 4):4–10.
13 Extra JM, Rousseau F, Bruno R et al: Phase I and pharmacokinetic study of Taxotere (RP 56976; NSC 628503) given as a short intravenous infusion. Cancer Res 1993;53:1037–1042.
14 Burris H, Irvin R, Kuhn J et al: Phase I clinical trial of Taxotere administered as either a 20-hour or 6-hour intravenous infusion. J Clin Oncol 1993;11:950–958.
15 Bissett D, Setanoians A, Cassidy J et al: Phase I and pharmacokinetic study of Taxotere (RP 56976) administered as a 24-hour infusion. Cancer Res 1993;53:523–527.
16 Pazdur R, Newman RA, Newman, et al: Phase I trial of Taxotere: Five-day schedule. J Natl Cancer Inst 1992;84:1781–1788.
17 Tomiak E, Piccart M, Kerger S et al: Phase I study of docetaxel administered as 1 hour intravenous infusion on a weekly basis. J Clin Oncol 1994;12:1458–1467.
18 Aapro MS, Zulian G, Alberto P et al: Phase I and pharmacokinetic study of RP 56976 in a new ethanol-free formulation of Taxotere. Ann Oncol 1992;3(suppl 5)53(abstr 208).
19 Pazdur R, Kudelka AP, Kavanagh JJ: Phase I studies of docetaxel (Taxotere) and selected phase II studies. Cancer Invest 1993;12(suppl 1):33–35.

20 Trudeau ME, Eisenhauer E, Lofters W et al: Phase II study of Taxotere as first line chemotherapy for metastatic breast cancer. A National Cancer Institute of Canada Clinical Trials Group Study (abstract). Proc Am Soc Clin Oncol 1993;12:64.

21 Piccart MJ, Klijn J, Paridaens R et al: Steroids do reduce the severity and delay the onset of docetaxel induced fluid retention: Final results of a randomized trial of the EORTC Investigational Drug Branch for Breast Cancer (abstract). Eur J Cancer 1995;31A(suppl 5):75.

22 Bruno R, Sanderink GJ: Pharmacokinetics and metabolism of Taxotere (Docetaxel); in Workman P, Graham MA (eds): Cancer Surveys: Pharmacokinetics and Cancer Chemotherapy. Cold Spring Harbor Laboratory Press, 1993, vol 17, pp 305–313.

23 Marlard M, Gaillard C, Sanderink G et al: Kinetics, distribution, metabolism and excretion of radiolabelled Taxotere in mice and dogs (abstract). Proc Am Assoc Cancer Res 1993;34:393.

24 Bissery MC, Bourzat JD, Commercon A et al: Isolation, identification, synthesis and biological activities of docetaxel metabolites. 207th American Chemical Society National Meetings, 13–17 March 1994, San Diego, Calif, abstr MEDI 144.

25 Vuilhorgne M, Gaillard C, Sanderink GJ et al: Metabolism of taxoid drugs; in Georg GI, Chen TT, Ojima I, Vyas DM (eds): Taxane Anticancer Agents: Basic Science and Current Status. ACS Symp Ser 1995;583:98–110.

26 Leonard RC, O'Brien M, Barrett-Lee P et al: A prospective analysis of 390 advanced breast cancer patients treated with Taxotere throughout the UK (abstract). Ann Oncol 1996;7(suppl 5):19.

27 Trudeau M: Docetaxel (Taxotere): An overview of first-line monotherapy. Semin Oncol 1995; 22(suppl 13):17–21.

28 Marre F, Sanderink GJ, de Sousa G et al: Hepatic biotransformation of docetaxel (Taxotere) in vitro: Involvement of the CYP3A subfamily in humans. Cancer Res 1996;56:1296–1302.

29 Fossella FB, Lee JS, Murphy WK et al: Phase II study of docetaxel for recurrent or metastatic non-small-cell lung cancer. J Clin Oncol 1994;12:1238–1244.

30 Cerny T, Kaplan S, Pavlidis N et al: Docetaxel (Taxotere) is active in non-small cell lung cancer: A phase II trial of the EORTC Early Clinical Trials Group. Br J Cancer 1994;70:384–387.

31 Francis PA, Rigas JR, Kris MG et al: Phase II trial of docetaxel in patients with stage III and IV non-small-cell lung cancer. J Clin Oncol 1994;12:1232–1237.

32 Miller VA, Rigas JR, Francis PA et al: Phase II trial of a 75 mg/m^2 dose of docetaxel with prednisone premedication for patients with advanced non-small-cell lung cancer. Cancer 1995;75:968– 972.

33 Kudo S, Hino M, Fukita A et al: Late phase II clinical study of RP 56976 (docetaxel) in patients with non-small-cell lung cancer. Jpn J Cancer Chemother 1994;21:2617–2623.

34 Chang AY, Kim K, Glick J et al: Phase II study of Taxol, merbarone and piroxantrone in stage IV non-small-cell lung cancer: The Eastern Cooperative Oncology Group results. J Natl Cancer Inst 1993;85:388–394.

35 Murphy WK, Fossella FV, Winn RJ et al: Phase II study of Taxol in patients with untreated advanced non-small-cell lung cancer. J Natl Cancer Inst 1993;85:384–388.

36 Fukuoka M, Fiitani H, Suzuki A et al: A phase II study of CPT-11, a new derivative of camptothecin, for previously untreated non-small cell lung cancer. J Clin Oncol 1992;10:16–20.

37 Negoro S, Fukuoka M, Niitani H et al: A phase II study of CPT-11, a camptothecin derivative, in patients with primary lung cancer. Jpn J Cancer Chemother 1991;18:1013–1019.

38 Douillard JY, Ibrahim N, Riviere A et al: Phase II study of CPT-11 (irinotecan) in non-small cell lung cancer (abstract). Proc Am Soc Clin Oncol 1995;14:365.

39 Abratt RP, Bezwoda WR, Falkson G et al: Efficacy and safety profile of gemcitabine in non-small-cell lung cancer: A phase II study. J Clin Oncol 1994;12:1535–1540.

40 Anderson H, Lund B, Bach B et al: Single-agent activity of weekly gemcitabine in advanced non-small lung cancer: A phase II study. J Clin Oncol 1994;12:1821–1826.

41 Gatzemeier U, Shepherd FA, Le Chevalier T et al: Activity of gemcitabine in patients with non-small cell lung cancer: A multicentre, extended phase II study. Eur Cancer 1996;32A:243– 248.

42 Depierre A, Lemarie E, Dabouis G, Garnier G, Jacoulet P, Dalphin JC: A phase II study of navelbine in the treatment of non-small cell lung cancer. Am J Oncol 1991;14:115–119.

43   Furuse K, Jubota K, Kawahara M et al: A phase II study of vinorelbine, a new derivative of vinca alkaloid, for previously untreated advanced non-small cell lung cancer. Lung Cancer 1994;11: 385–391.

44   Crivellari D, Veronesi A, Sacco C et al: Phase II study of vinorelbine in 50 patients with non-small cell lung cancer (abstract). Proc Am Soc Clin Oncol 1994;13:A1192.

45   Fossella FB, Lee JS, Shim DM et al: Phase II study of docetaxel for advanced or metastatic platinum-refractory non-small cell lung cancer. J Clin Oncol 1995;13:645–651.

46   Robinet G, Thomas P, Perol M et al: Phase II study of Taxotere (docetaxel) in advanced or metastatic non-small cell lung cancer previously treated with platinum. Ann Oncol 1996;7(suppl 5):96(abstr 458P).

47   Murphy WK, Winn WJ, Huber M et al: Phase II study of taxol in patients with non-small cell lung cancer who have failed platinum containing chemotherapy. Proc Am Soc Clin Oncol 1994;13: 363.

48   Bonomi PD, Finkelstein DM, Ruckdeschel JC et al: Combination chemotherapy versus single agents followed by combination therapy in stage IV non-small cell lung cancer: A study of the Eastern Cooperative Oncology Group. J Clin Oncol 1989;7:1602–1613.

49   Millward MJ, Zalcberg J, Bishop JF et al: Phase I trial of docetaxel cisplatin in previously untreated patients with advanced non-small cell lung cancer. J Clin Oncol 1996;15:750–758.

50   Verweij J, Planting AST, Vanderburg MEL et al: A phase I study of docetaxel and cisplatin in patients with solid tumours (abstract). Proc 8th NCI-EORTC New Drugs Symposium 1994, abstr 504, p 201.

51   Cole JT, Gralla RJ, Marques CB, Rittenberg CN: Phase I–II study of cisplatin + docetaxel (Taxotere) in non-small lung cancer. Proc Am Soc Clin Oncol 1995;15:351(abstr 1087).

52   Zalcberg J, Bishop JF, Millward MJ et al: Preliminary results of the first phase II trial of docetaxel in combination with cisplatin in patients with metastatic or locally adavnced non-small cell lung cancer (abstract). Proc Am Soc Clin Oncol 1995;14:351.

53   LeChevalier T, Belli L, Monnier A et al: Phase II study of docetaxel and cisplatin in advanced non-small cell lung cancer: An interim analysis (abstract). Proc Am Soc Clin Oncol 1995;14:350.

54   Bonomi P, Kim K, Chang A et al: Phase III trial comparing etoposide cisplatin versus taxol with cisplatin G-CSF versus taxol cisplatin in advanced non-small cell lung cancer. An Eastern Cooperative Oncology Group trial (abstract). Proc Am Soc Clin Oncol 1996;15:382.

55   Giaccone G, Splinter T, Postmus P et al: Paclitaxel cisplatin versus teniposide cisplatin in advanced non-small cell lung cancer (abstract). Proc Am Soc Clin Oncol 1996;15:373.

56   Johnson SA, Harper P, Hortolbagyi GN, Pouillart P: Vinorelbine: An overview. Cancer Treat Rev 1996;22:127–142.

57   Depieer A, Chastang C, Quoix E et al: Vinorelbine versus vinorelbine plus cisplatin in advanced non-small cell lung cancer: A randomised trial. Ann Oncol 1994;5:37–42.

58   LeChevalier T, Brisgand D, Douillard JY et al: Randomised study of vinorelbine and cisplatin versus vindesine and cisplatin versus vinorelbine alone in advanced non-small cell lung cancer: Results of a European multicentre trial including 612 patients. J Clin Oncol 1994;12:360–367.

59   Wozniak AJ, Crowly JJ, Balcerzak SP et al: Randomised phase II trial of cisplatin vs. cisplatin plus navelbine in the treatment of non-small cell lung cancer: A report of the South West Oncology Group (abstract). Proc Am Soc Clin Oncol 1996;15:374.

60   Kourousis C, Kakolyris S, Androullakis N et al: First-line treatment of non-small cell lung cancer with docetaxel and vinorelbine (abstract). Proc Am Soc Clin Oncol 1996;15:405.

61   Early E, Miller VA, Grant SC et al: Phase I/II trial of docetaxel and vinorelbine with filgrastrim in patients with advanced non-small cell lung cancer (abstract). Ann Oncol 1996;7(suppl 5):92.

62   Viullet J, Laberge F, Martins H et al: A phase II trial of docetaxel alternating with cisplatin and vinorelbine in non-small cell lung cancer (abstract). Ann Oncol 1996;7(suppl 5):93.

63   Dillman RO, Stephen L, Seagren M, Propert R et al: A randomised trial of induction chemotherapy plus high-dose radiation versus radiation alone in stage III non-small cell lung cancer. N Engl J Med 1990;323:940–945.

64   LeChevalier T, Arriaganda R, Tarayre M et al: Significant effect of adjuvant chemotherapy on survival in locally advanced non-small cell lung carcinoma (letter). J Natl Cancer inst 1992;84:58.

65 Sause W, Scott C, Taylor S et al: RTOG 8808, ECOG 4588. Preliminary analysis of a phase III trial in regionally advanced unresectable non-small cell lung cancer (abstract). Proc Am Soc Clin Oncol 1995;13:325.

66 Schaake-Konig C, van den Bogaert W, Dalesio O et al: Effects of concomitant cisplatin and radiotherapy on inoperable non-small cell lung cancer. N Engl J Med 1992;326:524–530.

67 Joschko MA, Webster LJ, Bishop JF et al: Radiopotentiation by paclitaxel in a human squamous carcinoma xenograft in nude mice. Radiat Oncol Invest 1996;4:268–274.

68 Masters GA, Haraf DJ, Hoffman PC, et al: Phase I study of concomitant docetaxel (Taxotere) and radiation in advanced chest malignancies. Proc Am Soc Clin Oncol 1996;15:314(abstr 1112).

Prof. James F. Bishop, Sydney Cancer Centre, Royal Prince Alfred Hospital,
Missenden Road, Camperdown, Sydney, NSW 2050 (Australia)
Tel. (2) 9515 7403, Fax (2) 9515 7404, E-Mail jbishop@canc.rpa.cs.nsw.gov.an

Schiller JH (ed): Updates in Advances in Lung Cancer. Prog Respir Res.
Basel, Karger, 1997, vol 29, pp 117–134

Chapter 8

·······················

The Roles of Radiation Therapy in the Treatment of Patients with Small Cell Lung Cancer: When, Where, and How Much Is Enough?[1]

Henry Wagner, Jr.

Thoracic Oncology Program, H. Lee Moffitt Cancer Center and Research Institute,
Tampa, Fla., USA

Introduction

Small cell lung cancer (SCLC) remains a major cause of death in the United States and the rest of the world [1]. Caused primarily by carcinogens in tobacco smoke, its incidence has risen steadily through this century, following changes in smoking behavior with about a 20- to 30-year lag. Thus, even if all smokers quit tomorrow, several million additional cases of cancer and deaths from SCLC can be expected during the next several decades. The development of more effective ways to diagnose and treat this disease has great value, even while our first priority is and should be its prevention.

Recognition of SCLC as a distinct clinical entity with both biologic and clinical properties different from those of other histologies of lung cancer began more than 40 years ago and was accelerated by the development of techniques for growth of pure cell lines in the 1970s [2, 3]. In comparison to other histologies of lung cancer (squamous cell, adenocarcinoma, large cell), SCLC was more likely to present with metastases to both regional lymph nodes and distant organs, have a more rapid clinical course (and clinical doubling time of measurable lesions), and a greater responsiveness to radiation

---

[1] Adapted from a presentation given at the IASLC Workshop on 'Controversies on Staging and Combined Modality Treatment of Lung Cancer' in Bruges, Belgium, June 1996, and Wagner H: Prophylactic cranial irradiation for patients with small cell lung cancer: An enduring controversy. Chest Surg Clin North Am 1997;7:151–166.

and to a variety of chemotherapeutic agents. For a time these observations led to a belief that only systemic therapy was needed for this disease and there was little or no place for radiation therapy (RT). During the past decade, however, there has been a return to the recognition, for lung cancer of all histologies, that the great majority of patients, even those with apparently localized disease, have systemic micrometastases, and that no patient is cured or long palliated without control of local intrathoracic disease. Thus the discussions have shifted from ones about the need for local therapy in SCLC (or systemic therapy in NSCLC) to those about the proper mode in which to combine local and systemic therapies [4–6].

Thoracic Radiation Therapy Improves Local Control and Survival

Many small to moderate-sized trials were conducted in the 1970s and 1980s comparing chemotherapy alone to chemotherapy plus thoracic radiation therapy (TRT) for patients with limited SCLC. These varied greatly in the choice of agents, dose, and duration of chemotherapy, the radiation target volume, dose, fractionation, and technique, and the relative timing of these two modalities. Some of these trials showed improved local control and survival for TRT, others did not. This left little consensus until the publication of two meta-analyses which combined the results of all published randomized trials comparing chemotherapy with or without TRT and showed a modest but statistically significant improvement in local control and survival [7–9]. As expected, toxicity, both hematologic and other, was greater with the addition of TRT but this was outweighed by the survival advantage. In the analysis reported by Pignon et al. [9], which used individual patient data, survival at 3 years was $8.9 \pm 0.9\%$ for chemotherapy alone compared to $14.3 \pm 1.1\%$ for chemotherapy plus TRT, a 33% relative difference.

These meta-analyses were neither intended nor able to detect differences between different modes (concurrent vs. sequential, early vs. late RT) of combining radiation and chemotherapy, or between different radiation doses or chemotherapeutic regimens (table 1). This task properly falls to randomized trials in which these questions can be isolated from other variables. The remainder of this section will explore these issues.

Radiation Target Volume

The appropriate radiation target volume for patients with limited SCLC has not been defined. Several distinct issues may be identified: the need for

Table 1. Sequencing and timing options in combining radiation and chemotherapy

---

Sequential
    CT→RT
    RT→CT

Alternating
    e.g. CT→RT→CT→RT→CT→RT→CT

Concurrent
    e.g. CT and RT given during the same time period and often on the same day. If concurrent
    treatment is given and the overall duration of RT is less than that of CT (which is almost
    always the case), several timing options are possible:
    Early CT/RT→CT→CT→CT
    Mid CT→CT→CT/RT→CT
    Late CT→CT→CT→CT/RT

---

margins of radiographically normal lung around a tumor mass, the amount of coverage for ill-defined tumor masses with associated atelectasis and/or infection, 'elective' coverage of regional lymph nodes, and the choice of treating the initial tumor volume or reduced postchemotherapy volume in regimens which defer TRT until after several cycles of chemotherapy.

No prospective trials have addressed the questions of margins, ill-defined tumors, or elective nodal volume. Older trials often recommended that portals cover all known primary disease with 2-cm margins and that elective nodal coverage includes bilateral hilar, mediastinal, and supraclavicular nodes. Retrospective analysis of patients on a SECSG trial showed that those not adhering to these requirements had a higher rate of local failure (69 vs. 32%) than those who did [10]. However, this analysis did not specify that the excess failure actually occurred in the areas not irradiated, nor did it consider that those patients treated not in conformity with the protocol may have been those with larger tumors whose portals were already quite large without addition of elective nodal irradiation. It was also conducted long before the routine use of CT scans in determining and planning radiation target volumes and has little applicability to current treatment techniques. More recent trials which have eliminated such elective irradiation of the contralateral hilum or any supraclavicular nodes, except in patients with known superior mediastinal disease, have not reported high failure rates in these regions [11].

If one adopts a treatment plan of deferring TRT until after several cycles of induction chemotherapy, a second question arises as to the appropriate radiation target volume. Should one target the initial (prechemotherapy) tumor volume or the reduced volume appearing following induction chemotherapy? This question has been formally addressed in only one prospective trial, done by the SWOG, in which patients receiving late TRT after achieving a partial response to initial chemotherapy were randomized between treatment to the initial or the postchemotherapy volume [25]. No benefit was seen for treating the larger volume, but this sequence of combining radiation and chemotherapy is one in which it has been difficult to see improvement by including TRT for any patients, and this randomization included only those patients having less than a CR to initial chemotherapy, so the sensitivity of this trial to detect a possible difference was low and its results do not apply at all to patients having a CR to induction chemotherapy.

Liengswangswong et al. [12] at the Mayo Clinic recently reviewed their limited SCLC experience. During the past decade they treated patients with several cycles of initial chemotherapy followed by TRT to the postchemotherapy target volume. They compared survival and local control in patients so treated to earlier patients who had RT to the initial volume, and found no differences. While their overall data were impressive, this study was rather small and retrospective, and the lack of demonstration of a benefit for treatment of the initial tumor volume should be viewed with considerable caution. Arriagada et al. [13] also found on retrospective review that patients treated to portals which were reduced to follow regression of tumor during alternating cycles of chemotherapy and TRT did not have inferior local control compared to those treated throughout to the initial tumor volume. In both of these studies the majority of failures were at the center of the target volume. If higher doses are able to achieve better control of central disease the issue of field margins may take on greater importance. A reduced target volume might also allow the use of higher total doses for the same level of acute and late toxicity, with possible improvement in local control. Treatment volume may well be an issue worth addressing prospectively, however, since both acute and late toxicities of TRT are volume dependent, and if treatment outcome is not jeopardized by a reduction in the target volume, as well as the delay in start of TRT needed to achieve this, there might well be gains in quality of life. Such a trial would be complex to conduct, requiring three arms (early TRT to the initial tumor volume, delayed TRT to the initial tumor volume, and delayed TRT to the postchemotherapy volume) to unambiguously resolve both the timing and volume issues, and a large sample size to detect or exclude small differences.

Optimal Radiation Dose and Fractionation Are Not Known

The observation that SCLC responded both more rapidly and frequently to radiotherapy than other histologies of lung cancer led to a belief that it

Table 2. Effect of radiation dose on local control in limited SCLC

Series	Drug regimen	Radiation dose	LC (1 year)	LC (2–3 years)	LC (NS)
IG-R [16]	MACE	45S	–	–	57
	PACE	55S	–	–	61
	PACE	65S	–	–	53
MGH [15]	COP-CAV	30C	~65	0 (3 years)	–
	COP-CAV	35C	~70	21 (3 years)	–
	COP-CAV	40C	~78	57 (3 years)	–
	MACC, VCE-VCA, PCE-ACE	45C	~80	61 (3 years)	–
	MACC, VCE-VCA, PCE-ACE	50C	~85	64 (3 years)	–
NCIC [17]	CAV, PE	25/10 fx		20 (2 years)	
	CAV, PE	37.5/15 fx		31 (2 years)	

could be controlled by relatively modest doses. Common treatment regimens in the 1970s gave doses in the range of 30 Gy/10 fractions. Since overall survival was relatively short and dominated by distant disease progression, it took some time to realize that such regimens were quite poor in achieving durable local control.

The inter-relation between radiation and chemotherapy in achieving and reporting local control is complex. Lacking effective chemotherapy, most patients die within 1 year, usually of systemic disease, and may not live long enough for locoregional relapse to become manifest. Survival is poor, but local control appears good. With better systemic chemotherapy, and survival and thus time at risk for local failure lengthend, the observed rate of locoregional relapse may increase unless the chemotherapy is also effective in controlling local disease at least as well as systemic disease. Arriagada et al. [14] clearly discussed the need for actuarial calculation of rates of relapse in local as well as systemic sites to better understand these relationships.

Recent series report that locoregional relapse occurs in about 50% of patients treated with tumor doses in the range of 40–50 Gy using conventional daily fractions of 1.8–2.5 Gy (table 2). It is not clear whether there is a disadvantage to split-course irradiation as is usually the case in the absence of effective chemotherapy. It is also not clear that dose escalation with single daily fractionation and thus lengthening of the overall treatment period is very effective in improving control (table 2). Choi and Carey [15] reported data from several sequential treatment regimens at the Massachusetts General Hospital and showed improvement in local control as doses increased from 30 to 50 Gy. Even at this higher dose, local failure at 2 years was seen in 40% of patients. Data from the Institute Gustav-Roussy [16] show similar local control for 45, 55 and 65 Gy (all split-course interdigitated with chemotherapy). The one trial which randomized patients to two dose levels was reported by Coy et al. [17] for the National Cancer Institute of Canada (NCIC). In this trial, patients were randomized between two relatively low-dose arms, 25 Gy/10 fractions and 37.5 Gy/15 fractions. The higher dose showed improved local control at early time points (2 years) but this decreased with time, indicating the inadequacy of both of these dose levels in producing durable local control.

Radiation Fractionation

The seeming failure of conventional dose escalation strategies in SCLC led investigators to explore alternate approaches to improve the effectiveness of local RT. One such approach is to deliver the total radiation dose in a shortened overall time by using multiple fractions per treatment day. Such an approach is particularly appealing for SCLC because of its rapid growth, which would argue for shortening the overall treatment time, and its relative lack of a shoulder on the radiation dose-survival curve when compared to most other tumor or normal tissues [18]. Such lack of a shoulder implies that a reduction in the size of each radiation fraction will have relatively less sparing of the SCLC tumor cells than of normal tissues, thus increasing the therapeutic ratio. The combination of twice-daily RT concurrent with cisplatin/etoposide chemotherapy was first introduced by Turrisi et al. [19] at the University of Pennsylvania and has since been tested in several variations (alternating rather than concurrent radiation and chemotherapy, delaying RT until after several chemotherapy cycles have been given, and giving sequential rather than concurrent chemotherapy and radiotherapy). With the exception of the regimen used at Memorial Sloan-Kettering Cancer Center, in which TRT was both delayed and given during a 3-week break from chemotherapy, results of the other series which include both concurrent and alternating sequences, as well as immediate TRT or after several initial cycles of chemotherapy, appear to give similar results [20].

A recent North American Intergroup (ECOG, RTOG, SWOG) trial has prospectively compared daily with twice-daily fractionation in patients with LSCLC [21]. The two fractionation schemes were 45 Gy/25 fractions/5 weeks

Table 3. North American Intergroup 0096 Phase III Trial of daily vs. BID TRT with concurrent cisplatin/etoposide in limited SCLC [from 21]

Arm	Patients	Resp (CR/CR + PR)	MST months	3-Year survival %	ESO TOX Grade 3–4
45 Gy/25 fractions/5 weeks	186	46.0/81.2	18.6	26.9	16
45 Gy/30 fractions/3 weeks	195	52.7/81.8	22.7	30.9	31[1]

[1] Grade 4 esophageal toxicity was 5% for both arms, the difference between the two was the incidence of grade 3. All other nonhematologic and hematologic toxicities were the same on both arms.

or 45 Gy/30 fractions/3 weeks. Both were begun concurrently with the first of four cycles of cisplatin and etoposide. Present results show excellent median and 3-year survival for both arms with a trend (not statistically significant) to better time to progression and survival for the twice-daily regimen (table 3). Local control was significantly improved with twice-daily TRT. While this regimen was associated with more grade 3 esophagitis (25.7 vs. 10.9%) other nonhematologic and hematologic toxicities were identical.

Improving Local Control

Current approaches to improving local control have taken several directions. Choi et al. [22] have explored dose escalation with both daily (2 Gy) and BID (1.5 Gy) fractionation in a sequential cohort dose-seeking phase I study of 42 patients. RT was given following three cycles of induction chemotherapy with cisplatin/etoposide/cyclophosphamide and with cisplatin/etoposide given concurrently with radiation. Dose-limiting (grade 3) acute esophagitis was reached by their criteria at total doses of 45 Gy for the BID regimens and 70 Gy for the daily regimen. No data have been reported on late esophageal strictures or on local control or survival, which will require much larger groups of patients for reasonable estimates of these parameters.

In addition to increasing total radiation dose, attempts are being made to improve local control by combining radiation with low-dose weekly or daily chemotherapy, hypoxic radiation sensitizers such as the nitroimidazoles, selective cytotoxins of hypoxic cells such as tirapazamine. Another approach which takes advantage of the near-universal lack of normal p53 gene function in SCLC lines is to combine radiation with the administration of a drug, such as lysofylline, which preferentially radiosensitizes cells lacking normal G1 checkpoint function [23].

Sequencing and Timing of Radiation and Chemotherapy

TRT and chemotherapy may be combined in many permutations (table 1). These involve choices of both timing (will radiation be given before, during, or after chemotherapy) and concurrence (will radiation and chemotherapy be given simultaneously or sequentially). There are both theoretic and logistic factors involved in the choice of timing and concurrence. The use of both modalities simultaneously increases acute toxicities, particularly esophagitis, and to some degree myelosuppression. This is less of a problem with chemotherapeutic regimens now in common use such as cisplatin/etoposide (PE) compared with earlier ones such as cyclophosphamide/doxorubicin/vincristine (CDV), but can be an issue for some patients, particularly those of more advanced age, borderline performance status, or gastroesophageal reflux. Starting radiation therapy after several cycles of chemotherapy may allow treating a smaller target volume if the tumor regresses substantially during chemotherapy, but poses the risk of allowing both proliferation of cell clones which may be radiation sensitive but drug resistant.

Few published trials speak directly to the issues of the timing and sequencing of radiation and chemotherapy in limited (LSCLC). Those which have been reported vary in important details and it is not surprising that they reach different conclusions (table 4).

The Cancer and Leukemia Group B (CALGB) conducted a three-arm trial in which patients were randomized among chemotherapy alone or combined with TRT (50 Gy/6 weeks) begun either on day 1 or day 64 of chemotherapy and given concurrently [24]. The arms including TRT were superior to the chemotherapy-only regimen for both local control and survival, and there was a trend favoring the delayed TRT arm (2-year survival 25 vs. 15%). At recent update the 5-year survival continues to favor delayed TRT, 3% for chemotherapy alone, 6.6% for early TRT and 12.8% for delayed TRT [25]. However, in this trial the dose intensity of chemotherapy (using a noncisplatin-based regimen) was deliberately reduced in the early TRT arm, and there are some data that such early drug dose intensity is an important determinant of survival.

The NCIC randomized patients receiving alternating cycles of CDV and EP to receive TRT (40 Gy/15 fractions/3 weeks) concurrent with either the first or third cycle of PE [26]. Survival was significantly superior for patients receiving early TRT (median 21.2 vs. 16 months, 4-year 25 vs. 15%). Somewhat

Table 4. Randomized trials of radiation-chemotherapy sequencing and timing in LSCLC

Trial	Comparison	Chemotherapy	Result	Comment
Perry [25]	TRT d1 vs. d64 vs. none	CDE/CEV	Delayed TRT gave best survival	Planned drug dose reduction in arm with d1 TRT
Murray [26]	TRT cycle 2 vs. cycle 6	CDV/PE alternating	Median and 4-year survival favor early TRT	Greatest difference in patterns of failure was in CNS relapse
Schultz [27]	TRT d1 vs. d120	CDV/PE alternating	Delayed TRT nonsignificantly better (MST 12.9 vs. 10.7 months)	No long-term follow-up
Takada [28]	TRT concurrent with cycle 1 or after cycle 4	PE	MST significantly better for early concurrent rather than late sequential TRT (31.3 vs. 20.8 months)	Median follow-up only 18.6 months, await longer-term data
Gregor [29]	TRT alternating between cycles 2–3–4–5 or after cycle 5	CDE	Nonsignificant trends favoring delayed TRT	Alternating arm had more hematologic toxicity which led to drug dose reduction and TRT delay or incompletion
LeBeau [30]	TRT concurrent with cycle 2 or alternating between cycles 2–3–4–5	CDE	No difference in median survival	No long-term data; poorer compliance in alternating arm

surprisingly, the main difference in patterns of relapse was in the incidence of CNS metastases, not in control of intrathoracic disease.

Shultz et al. [27] treated patients with limited SCLC using alternating cycles of EP and CDV and randomized them between TRT (40 Gy/22 fractions/ $4\frac{1}{2}$ weeks or 45 Gy/22 fractions by split course) starting on day 1 or day 120. There was a nonsignificant difference in median survival favoring delayed TRT (10.7 vs. 12.9 months). Long-term survival and local control were reported. Both of the above trials used rather low total radiation doses.

Takada et al. [28] reported preliminary results of a phase III trial conducted by the Japanese Clinical Oncology Group in which patients with

LSCLC were treated with four cycles of cisplatin/etoposide chemotherapy and randomized to receive TRT (45 Gy/30 fractions/3 weeks) starting either concurrent with the first or following the fourth cycle of chemotherapy. With a median follow-up time of 18.6 months, median survival was 20.8 months for the delayed TRT and 31.3 months for early TRT. In interpreting these results, it should be kept in mind that two comparisons were made in this trial, early versus late TRT and concurrent versus sequential chemoradiation and that neither question on its own will be unambiguously answered by it.

Gregor [29] reported preliminary results of an EORTC/UKCCCR trial which compared alternating TRT and chemotherapy with late TRT. Chemotherapy was with cyclophosphamide/doxorubicin/etoposide. RT was given either interspersed between cycles of chemotherapy (1,000 cGy/4 fractions/ 1 week) for four courses or at completion of all chemotherapy. Survival at 3 years was not significantly different (12% for alternating vs. 15% for sequential treatment). Acute hematologic toxicity was much worse in the alternating arm. This resulted in significantly lower drug dose intensity (by about 25%) in patients treated on the alternating arm. RT intensity was also reduced, with RT being completed in 93% of patients receiving late RT compared to only 77% completion for alternating RT. This trial highlights some of the difficulties of trying to answer questions about optimal RT scheduling outside the context of the chemotherapy used. The CDE regimen which was used in this study is more myelosuppressive than PE and conclusions which apply to the combination of RT with one may not apply to the other.

LeBeau et al. [30] reported a randomized trial comparing concurrent RT starting with the second cycle of chemotherapy (50 Gy/20 fractions) with alternating treatment between cycles 2–3–4–5. In the alternating regimen the radiation dose for the first two courses was 20 Gy/8 fractions and was 15 Gy/ 6 fractions for the third course. Chemotherapy was CDE in both arms. Median survival was 407 days for the concurrent arm and 426 for the alternated. More severe lung fibrosis was noted with concurrent therapy, and compliance with the schedule of radiotherapy was reported to be poor, particularly for the alternating arm.

In an attempt to summarize these results of various trials which combined radiation and chemotherapy for SCLC, Murray et al. [31] analyzed 3-year survival as a function of the interval from the start of chemotherapy to the start of RT. There was significant superiority for those regimens beginning RT not more than 6 weeks after the start of chemotherapy (table 5). Delay of the start to TRT to 20 weeks or more gave results not much better than with chemotherapy alone.

Table 5. Meta-analysis of timing of thoracic irradiation relative to chemotherapy in LSCLC [from 31]

Interval, weeks	Mean interval, weeks	Patients	3-Year PFS, %
0–2	0	426	18.9
3–5	4	304	22.2
6–10	9	376	14.1
11–19	17	453	12.7
20+	20	388	13
Never	n/a	493	6.7

PFS = Progression-free survival.

Prophylactic Cranial Irradiation

CNS involvement is common in SCLC. At diagnosis about 20% of patients will have demonstrable CNS metastases, usually in combination with other extrathoracic sites. Patients with limited SCLC treated effectively with thoracic radiation and chemotherapy have a time-dependent risk of developing overt CNS disease which reaches about 60–70% at 3 years [32]. Previously, with shorter average survivals and lack of actuarial reporting, this was often underestimated as a crude risk of 20–30%. Even with these lower estimates, however, it was recognized that CNS relapse was frequent enough to warrant attention. Hansen et al. [33] suggested that the CNS might be a pharmacologic sanctuary site and that prophylactic cranial irradiation (PCI) could be used to prevent these seemingly isolated relapses. Such an argument was based on the success of a similar strategy in pediatric acute lymphoblastic leukemia, in which prophylactic treatment of the craniospinal axis with craniospinal irradiation or cranial irradiation and intrathecal chemotherapy resulted in decrease of meningeal relapse and improved survival.

To be valid, this argument requires that: (1) some patients with SCLC will relapse only in the CNS; (2) prophylactic (or preemptive) treatment of occult CNS metastases is more effective than treatment of overt (either radiographically or clinically) disease, either in achieving long-term disease control or maintaining quality of life, and (3) prophylactic treatment has acceptable toxicity.

Several randomized trials have tried to determine the role of PCI in patients with SCLC (table 6). Many were flawed by the inclusion of patients without a good response to initial therapy (either a CR or major PR), did not report patterns of relapse (e.g. did patients who failed in the brain have

Table 6. Randomized prospective trials of PCI in SCLC-CNS relapse rates and survival

Table 6. Randomized prospective trials of PCI in SCLC-CNS relapse rates and survival

Series	Patients	% CNS failure		Median survival months		2-year survival, %	
		PCI	no PCI	PCI	no PCI	PCI	no PCI
Jackson [49]	29	0	27	9.8	7.2	8.6	17
Cox [50]	45	17	24	NR	NR	NR	NR
Maurer [51]	163	4	18	8.4	8.8	8.3	13.8
Hansen [52]	109	9	13	9.4	10.3	11	17
Aisner [53]	29	0	36	NR	NR	NR	NR
Niiranen [54]	51	0	27	13	10	NR	NR
Beiler [55]	54	0	16	7.8	8.0	NR	NR
Eagan [56]	30	13	73	13.8	12.9	NR	NR
Seydel [57]	219	5	20	10.3	10.3	NR	NR
Arriagada [34]	145	40[1]	67	~14	~12	21.5	29
Arriagada [34]		19[2]	45				
Gregor [35]	314	HR 0.41[3]	NR	NR	NR	NR	NR
Wagner [36]	31	20	50	15.3[4]	8.8[4]	NR	NR

NR = Not reported.
[1] As any site of failure.
[2] As first and isolated site of failure.
[3] Hazard ratios compared with patients not receiving PCI.
[4] Survival from PCI randomization, which was a median of 6 months from diagnosis.

isolated relapse or fail in multiple sites either then or later?) or did not prospectively follow patients for disease or treatment-related neurologic dysfunction. More recent trials have attempted to remedy these deficiencies.

Recent Prospective Trials

Arriagada et al. [34] reported preliminary results of a randomized trial which compared PCI (25 Gy/10 fractions) to observation in patients with SCLC who had achieved CR to induction treatment using chemotherapy with or without TRT. Three hundred patients were randomized over 7 years, 240 of them with limited disease. At 2 years after randomization, overall rates of brain metastases (40 vs. 67%) and isolated brain metastases (19 vs. 45%) were significantly lower in patients receiving PCI. Overall survival was 29.5 vs. 21% (p = 0.14). There was a nonsignificant increase in the frequency of deterioration of intellectual function in patients receiving PCI as well as an almost significant difference in morphologic changes on CT scan (11 vs. 3%, p=0.06). Longer follow-up and a larger confirmatory trial are being conducted to clarify the suggestions of both survival benefit and effects on quality of life.

In North America, the ECOG and RTOG compared PCI (25 Gy/10 fractions/2 weeks) to close observation and therapeutic cranial irradiation in patients who developed brain metastases [35]. This trial was open to any patient with SCLC who had achieved a CR (or PR with minimal residual fibrosis), regardless of initial stage of disease or treatment used to reach CR. Patients were stratified by their pretreatment weight loss and the presence or absence of liver metastases. Both of these factors, but not such initially important factors such as disease stage, sex, or performance status, had been shown predictive of survival after achieving CR. This trial opened in 1989 and closed in 1993, having accrued only 31 patients of a planned 150. The primary endpoint of the trial was the rate of CNS relapse, and the sample size was chosen to demonstrate a 20% difference in rates. Because of the low accrual, conclusions from this trial are tentative. In the PCI arm there were 3/16 CNS relapses, all isolated, while in the observation arm there were 8 CNS relapses, 4 isolated and 4 in conjunction with other systemic sites of failure. This difference was statistically significant. Overall survival approached statistical significance (p=0.07) with 40% of PCI patients and 15% of observation patients alive at 2 years from the PCI randomization (about 2.5 years from diagnosis).

Gregor et al. [36] reported initial results of a UKCCCR/EORTC trial which randomized patients between PCI (with different institutions allowed to use their preferred dose and fractionation regimen) and observation. PCI significantly reduced the risk of developing brain metastases (hazard ratio 0.41, p<0.0001). A dose-response relation for prevention of CNS relapse was noted between 24 and 36 Gy. No significant survival difference has been seen between the PCI and no-PCI groups. Further analysis of survival as a function of PCI dose is ongoing, although this comparison was not protected by randomization since PCI dose was an institutional choice.

An international meta-analysis, using individual patient data, of all published randomized trials of PCI is currently being conducted and preliminary results should be available in the summer of 1997.

Do Patients Who Receive PCI Have Neurologic Deficits, and Does PCI Cause Them?

Several reports in the 1980s described a variable, and occasionally alarmingly high, incidence of neurologic dysfunction in patients who had received

PCI as part of their treatment [37–45]. This in conjunction with the uncertain survival benefits led to a backlash against its routine use. Subsequent analysis has suggested that the issue of neurologic dysfunction is complex, that multiple factors must be taken into account, and that dysfunction is not simply explained by either radiotherapeutic or chemotherapeutic toxicities [46]. Factors which must be considered include:

(1) What is the baseline neurologic function of the patient population? Patients with SCLC are often elderly smokers who have chronic pulmonary, cardiovascular, and cerebrovascular disease. They are often taking multiple medications which may affect cognitive functioning. Several recent studies have clearly shown that neuropsychiatric testing of individuals with SCLC reveals abnormalities in 30–40% prior to PCI, and have not shown consistent further deterioration following PCI [36, 39, 47, 48]. Using age- and sex-matched controls without considering smoking history and co-morbid diseases may not be appropriate in studying cognitive function in patients with SCLC.

(2) Are patients with SCLC more likely than those with NSCLC to have pretreatment neurologic dysfunction, possibly mediated through neurohumoral mechanisms or between antigenic cross-reactivity between tumor and neural antigens? Some such syndromes, e.g. retinopathy, cerebellar degeneration, are well described. Is neurologic dysfunction in some patients with SCLC due to paraneoplastic mechanisms?

(3) Radiation damage to the CNS certainly occurs and is generally dose and fractionation dependent. Some series which reported a high incidence of neurologic dysfunction following PCI used large daily doses (e.g. 3–4 Gy). More recent series have used smaller fraction sizes and report less toxicity; is this due to the reduced fraction size or shorter duration of follow-up?

(4) Neurotoxicity from chemotherapy, both on its own and in combination with radiation, must be considered. Several agents which were widely used to treat SCLC, including methotrexate, procarbazine, and the nitrosoureas, have good access to the CNS despite an intact 'blood-brain barrier' and appear to be associated with increased neurotoxicity. Furthermore, there is clear evidence that cranial irradiation increases permeability of the blood-brain barrier so that agents not normally reaching the CNS can do so. This would increase the potential for neurologic toxicity in patients receiving 'maintenance' following PCI.

(5) Several distinct entities have been reported in publications describing neurologic dysfunction. These range from radiologic changes seen on CT or MRI with or without clinical correlate, to abnormal results on psychometric testing in individuals functioning well in their daily activities, to clear clinical dementia preventing the patient from leading an enjoyable or productive life. While these are all of research they take quite differing weight in comparing

---

Wagner, Jr.   130 their risks to that of developing brain metastases, which are almost invariably fatal, and which, while amenable to at least short-term palliation in the majority of patients, will be symptomatic at the time of death in the majority of patients who develop them.

There is no present consensus on the role of PCI in patients with SCLC. On several points, however, most investigators would agree that: (1) PCI should be considered only for patients achieving a complete or near complete response to their initial treatment with chemotherapy +/− thoracic irradiation; (2) PCI results in a significant delay and reduction in the overall risk of brain metastases and isolated brain metastases, although with currently used doses this risk remains high, warranting consideration of study of higher doses, and (3) if PCI has an impact on overall survival, it is likely to be modest, in the range of 5–10%.

The greater controversies rest with whether the possible survival advantage has been convincingly demonstrated with current trials, whether the reduction in brain metastases even in patients who go on to die of active disease in other sites is of sufficient value in quality of life to be worthwhile, and the toxicities associated with PCI. At present none of these questions is definitively answered by prospective trials. My interpretation of the recent larger trials is that they do strongly suggest a modest survival gain consistent with what one might expect, that the survival and quality of life of patients relapsing in the brain is quite poor and minimally impacted by therapy at the time of relapse, and that the 'toxicity' attributed in the past to PCI is in reality a complex mixture of pre-existing neurologic impairment, effects of treatment (both radiation and chemotherapy) and depression. Weighing these factors, I remain a proponent of PCI in appropriate patients. I do feel, however, that patients should be advised of both the potential risks as well as modest benefits of PCI and assisted in making an informed choice consistent with their own preferences regarding risks of treatment versus risks of disease. PCI may not be a walk in the park but it beats developing brain metastases.

Conclusions

Limited SCLC is a curable disease in a minority of patients. All patients in good overall health (performance status, weight loss) should be treated aggressively with both radiation and chemotherapy. The newly diagnosed patient should be evaluated jointly by radiation and medical oncologists for such planned combined modality treatment. Radiation and chemotherapy should both be given early in the patient's treatment, either concurrently or closely sequenced. The toxicities of such combined therapies at present mandate aggressive supportive care through periods of anticipated myelosuppression and esophagitis but are clearly warranted in the curative setting. While further improvements in local control and prevention of CNS are needed, the greatest avenue for improvement in SCLC remains with better treatment for systemic disease.

References

1   Boring CC, Squires TS, Tong T: Cancer statistics 1993. CA Cancer J Clin 1993;43:7–26.
2   Pettengill OS, Faulkner CS, Wurster-Hill DH, et al: Isolation and characterization of a hormone-producing cell line from human small anaplastic carcinoma of the lung. J Natl Cancer Inst 1977; 58:511–517.
3   Carney DN, Gazdar AF, Bepler G, et al: Establishment and characterization of SCLC cell lines having classic and variant features. Cancer Res 1985;45:2913–2923.
4   Wagner H: Limited small cell lung cancer: Current treatment and clinical trials. Cancer Control 1994;1:485–497.
5   Arriagada R, LeChevalier T: Small cell lung cancer – Regional disease; in Green Lung Cancer Book, 1996, chapt 21, pp 456–495.
6   Williams TE, Turrisi AT III: Role of radiotherapy in the treatment of small cell lung carcinoma. Chest Surg Clin North Am 1997;7:135–149.
7   Warde P, Payne D: Does thoracic irradiation improve survival and local control in limited-stage small-cell carcinoma of the lung? J Clin Oncol 1992;10:890–895.
8   Pignon J-P, et al: A meta-analysis of thoracic radiotherapy for small-cell lung cancer. N Engl J Med 1992;327:1618–1624.
9   Pignon J-P, et al: A meta-analysis using individual patient data from randomized clinical trials of chemotherapy in non-small cell lung cancer. 2. Survival in the locally advanced setting (meeting abstract). Proc Annu Meet Am Soc Clin Oncol 1994;13.
10 Perez C, Krauss S, Bartolucci A, et al: Thoracic and elective brain irradiation with concomitant or delayed multiagent chemotherapy in the treatment of localized small cell carcinoma of the lung. Cancer 1981;47:2407–2413.
11 Turrisi AT III, Glover D, Mason B, et al: Long-term results of platinum etoposide + twice daily thoracic radiotherapy for limited small cell lung cancer: Results on 32 patients with minimum 48 month F/U. Proc Am Soc Clin Oncol 1992;11:292(abstr 975).
12 Liengswangswong V, Bonner J, Shaw E, et al: Limited stage small cell lung cancer: Patterns of recurrence and implications for thoracic radiotherapy. J Clin Oncol 1994;12:496–502.
13 Arriagada R, Pellae-Cosset B, Cueto Ladron de Guevera J, et al: Alternating radiotherapy and chemotherapy schedules in limited small cell lung cancer: Analysis of local chest recurrences. Radiother Oncol 1991;20:91–98.
14 Arriagada R, et al: Competing events determining relapse-free survival in limited small-cell lung carcinoma. J Clin Oncol 1992;10:447–451.
15 Choi N, Carey N: Importance of radiation dose in achieving improved locoregional tumor control in limited stage small-cell lung carcinoma: An update. Int J Radiat Oncol Biol Phys 1989;17:307–310.
16 Arriagada R, et al: Alternating radiotherapy and chemotherapy in limited small cell lung cancer: The IGR protocols. French FNCLCC Lung Cancer Study Group. Lung Cancer 1994;10.
17 Coy P, et al: Patterns of failure following loco-regional radiotherapy in the treatment of limited stage small cell lung cancer [see comments]. Int J Radiat Oncol Biol Phys 1994;28:355–362.
18 Carney D, Mitchell J, Kinsella T: In vitro radiation and chemotherapy sensitivity of established cell lines of human small cell lung cancer and its large cell morphological variant. Cancer Res 1983; 43:2806–2811.

19   Turrisi A, Glover D, Mason B: A preliminary report: Concurrent twice-daily radiotherapy plus platinum-etoposide chemotherapy for limited small cell lung cancer. Int J Radiat Oncol Biol Phys 1988;15:183–187.

20   Armstrong J, et al: Limited small cell lung cancer: Do favorable short-term results predict ultimate outcome? Am J Clin Oncol 1991;14:285–290.

21   Johnson D, et al: Cisplatin and etoposide + thoracic radiotherapy administered once or twice daily in limited stage small cell lung cancer: Final results of Intergroup Trial 0096. Proc Am Soc Clin Oncol 1996;15(abstr 1113).

22   Choi N, Herndon J, Rosenman J, et al: Phase I study to determine the maximum tolerated dose of radiation in standard daily and accelerated twice daily radiation schedules with concurrent chemotherapy for limited stage small cell lung cancer: CALGB 8837. Proc Am Soc Clin Oncol 1995;14:363(abstr 1113).

23   Russell KJ, Wiens LW, Demers GW, et al: Preferential radiosensitization of G1 checkpoint-deficient cells by methylxanthines. Int J Radiat Oncol Biol Phys 1996;36:1099–1106.

24   Perry M, et al: Chemotherapy with or without radiation therapy in limited small-cell carcinoma of the lung. N Engl J Med 1987;316:912–918.

25   Perry M, et al: Thoracic radiation therapy added to chemotherapy in limited small cell lung cancer: An update of Cancer and Leukemia Group B Study 8083. Proc Am Soc Clin Oncol 1996(abstr 1150).

26   Murray N, et al: Importance of timing for thoracic irradiation in the combined modality treatment of limited-stage small-cell lung cancer. J Clin Oncol 1993;11:336–344.

27   Schultz H, et al: Timing of chest irradiation with respect to combination chemotherapy in small cell lung cancer, limited disease. Lung Cancer 1988;4:153.

28   Takada M, et al: Phase III study of concurrent versus sequential thoracic radiotherapy in combination with cisplatin and etoposide for limited-stage small cell lung cancer: Preliminary results of the Japanese Clinical Oncology Group. Proc Am Soc Clin Oncol 1996;15(abstr 1103).

29   Gregor A: Phase III comparison of alternating and sequential radiation therapy and CDE chemotherapy for limited SCLC; in IASLC, Bruges, Belgium, 1996.

30   LeBeau B, et al: A randomized clinical trial comparing concurrent and alternated thoracic irradiation in limited small cell lung cancer. Proc Am Soc Clin Oncol 1996;15(abstr 1148).

31   Murray N, et al: Importance of timing for thoracic irradiation in the combined modality treatment of limited-stage small-cell lung cancer. J Clin Oncol 1993;11:336–344.

32   Arriagada R, et al: Randomized trial on prophylactic cranial irradiation for patients with small cell lung cancer in complete remission (meeting abstract). Proc Annu Meet Am Soc Clin Oncol 1994;13.

33   Hansen H, et al: Prophylactic irradiation in bronchogenic small cell anaplastic carcinoma: A comparative trial of localized versus extensive radiotherapy including prophylactic brain irradiation in patients receiving combination chemotherapy. Cancer 1980;46:279–284.

34   Arriagada R, et al: Prophylactic cranial irradiation for patients with small-cell lung cancer in complete remission. J Natl Cancer Inst 1995;87:183–190.

35   Wagner HJ, et al: A randomized phase III trial of prophylactic cranial irradiation vs. observation in patients with small cell lung cancer achieving a complete response: Final report of an incomplete trial by the Eastern Cooperative Oncology Group and Radiation Therapy Oncology Group (E3589/R92-01) (abstract). Proc Am Soc Clin Oncol 1996;15:1120.

36   Gregor A, et al: Effects of prophylactic cranial irradiation in small cell lung cancer; Results of UKCCCR/EORTC Randomised Trial (abstract). Proc Am Soc Clin Oncol 1996;15:1139.

37   Chak L, et al: Neurologic dysfunction in patients treated for small cell carcinoma of the lung: A clinical and radiological study. Int J Radiat Oncol Biol Phys 1986;12:385–389.

38   Crossen J, et al: Neurobehavioral sequelae of cranial irradiation in adults: A review of radiation-induced encephalopathy. J Clin Oncol 1994;12:627–642.

39   Cull A, et al: Neurological and cognitive impairment in long-term survivors of small cell lung cancer. Eur J Cancer 1994;30A:1067–1074.

40   Frytak S, et al: Leukoencephalopathy in patients with small cell lung cancer receiving prophylactic cranial irradiation. Proc Am Soc Clin Oncol 1987;686.

41   Johnson B, et al: Neurologic, neuropsychologic, and cranial computed tomography scan abnormalities in 2–10 year survivors of small cell lung cancer. J Clin Oncol 1985;3:1659–1667.

42   Johnson B, et al: Neurologic, computed cranial tomographic, and magnetic resonance imaging abnormalities in patients with small cell lung cancer: Further follow-up of 6- to 13-year survivors. J Clin Oncol 1990;8:46–56.

43   Lee J, et al: Neurotoxicity in long-term survivors of small cell lung cancer. Int J Radiat Oncol Biol Phys 1986;12:313–321.

44   Licciardello J, et al: Disturbing central nervous system complications following combination chemotherapy and prophylactic whole-brain irradiation in patients with small cell lung cancer. Cancer Treat Rep 1985;69:1429–1430.

45   Lishner M, et al: Late neurologic complications after prophylactic cranial irradiation in patients with small cell lung cancer: The Toronto experience. J Clin Oncol 1990;8:215–221.

46   Turrisi A, Brain irradiation and systemic chemotherapy for small-cell lung cancer: Dangerous liaisons? (editorial) J Clin Oncol 1990;8:196–199.

47   Komaki R, et al: Evaluation of cognitive function in patients with limited small cell lung cancer prior to and shortly following prophylactic cranial irradiation. Int J Radiat Oncol Biol Phys 1995; 33:179–182.

48   Van Oosterhout A, et al: Neurologic disorders in 203 consecutive patients with small cell lung cancer. Results of a longitudinal study. Cancer 1996;77:1434–1441.

49   Jackson D, Richards F, et al: Prophylactic cranial irradiation in small cell carcinoma of the lung. JAMA 1977;237:2730–2933.

50   Cox JD, Petrovicj Z, et al: Prophylactic cranial irradiation in patients with inoperable carcinoma of the lung. Cancer 1978;42:1135–1140.

51   Maurer L, Tulloh M, et al: A randomized combined modality trial in small cell carcinoma of the lung: Effects of maintenance chemotherapy and prophylactic whole brain irradiation. Cancer 1980; 45:30–39.

52   Hansen H, Dombernowski P, et al: Prophylactic irradiation in bronchogenic small cell anaplastic carcinoma: A comparative trial of localized versus extensive radiotherapy including prophylactic brain irradiation in patients receiving chemotherapy. Cancer 1980;46:279–284.

53   Aisner J, Whitacre M, et al: Combination chemotherapy for small cell carcinoma of the lung: Continuous versus alternating non-cross-resistant regimens. Cancer Treat Rep 1982;66:221–230.

54   Niiranen A, Holsti P, et al: Treatment of small cell lung cancer: Two-drug versus four-drug chemotherapy and loc-regional irradiation with or without prophylactic cranial irradiation. Acta Oncol 1989; 28:501–505.

55   Beiler D, Kane R, et al: Low dose elective brain irradiation for small cell carcinoma of the lung. Int J Radiat Oncol Biol Phys 1979;5:944–945.

56   Eagan R, Frytak S, et al: A case for preplanned thoracic and prophylactic brain radiation therapy in limited small cell lung cancer. Cancer Clin Trials 1981;4:261–266.

57   Seydel H, Creech R, et al: Prophylactic versus no brain irradiation in regional small cell lung carcinoma. Am J Clin Oncol 1985;8:218–223.

Henry Wagner, Jr., M.D., Program Leader, Thoracic Oncology Program,
H. Lee Moffitt Cancer Center and Research Institute, Tampa, FL 33612, Florida, USA

Schiller JH (ed): Updates in Advances in Lung Cancer. Prog Respir Res.
Basel, Karger, 1997, vol 29, pp 135–149

Chapter 9
························

New Drug Therapies for Small Cell Carcinoma of the Lung

Joan H. Schiller

Department of Medicine, Medical Oncology Section, University of Wisconsin
Comprehensive Cancer Center, Madison, Wisc., USA

Small cell lung cancer (SCLC) is a neoplasm which is highly sensitive to chemotherapy. Standard chemotherapy regimens, which usually consist of combinations of cisplatin and etoposide (PE) or cyclophosphamide/Adriamycin/vincristine (CAV), typically yield response rates of 70–80% in patients who present with limited disease, and 50–80% in patients with extensive-stage disease [1–4]. Despite these dramatic initial responses, patients soon relapse, so that median survival is measured in months. Clearly, better chemotherapy regimens are needed for this disease. This chapter reviews some of the newer agents or drug combinations available for both previously untreated and relapsed or refractory disease.

Given the high response rate to standard combination chemotherapy regimens, questions have been raised as to the safety and appropriateness of administering new drugs to patients with this highly responsive disease. However, given the overall dismal cure rate for these patients, it is imperative we identify better therapies. Phase II trials have demonstrated that it is safe and ethical to test new agents in patients with previously untreated SCLC, particularly when arrangements have been made (such as a cross-over design) for the patients to receive standard therapy in the event of progression or relapse [5, 6]. Current recommendations regarding drug development for SCLC indicates it is appropriate to test new agents, particularly when the drugs: (1) have novel or unknown mechanisms of activity and show clinical activity in other solid tumors; (2) show specific activity for SCLC in preclinical drug screens, or (3) are analogs of active drugs, and are being developed primarily because of a different toxicity profile [7].

Carboplatin

The combination of PE has been one of the standard treatments for SCLC, due to the in vitro synergy of the combination, the improved tolerability when combined with thoracic irradiation compared to CAV, and high response rates in patients who have failed primary therapy with CAV. However, the administration of PE can be associated with nausea and vomiting, renal toxicity, and neuropathy. Carboplatin, a platin analog with the principle toxicity of myelosuppression, also has significant single-agent activity in the disease. The Hellenic Cooperative Oncology Group compared PE with carboplatin and etoposide (CE) in previously untreated limited-stage or extensive-stage patients [8]. One hundred forty-seven patients were randomized to receive either 50 mg/m^2/day of cisplatin on days 1 and 2 or 300 mg/m^2 of carboplatin on day 1, both combined with 300 mg/m^2/day of etoposide for 3 days. Treatment was administered every 21 days for 6 cycles; responding limited-stage patients and completely responding extensive-stage patients also received thoracic irradiation and prophylactic cranial irradiation. The complete response rates were 57 and 58% for the PE and CE arms, respectively. Median survival was 12.5 months on the PE arm and 11.8 months on the CE arm. The PE arm was associated with more grade 2 nausea and vomiting (65 vs. 27%) and grade 2 neuropathy (8 vs. 1.4%) than CE; 2 of 73 patients died of neutropenic infections on the PE arm, compared to none of 74 patients randomized to receive CE. These data suggest that CE is as efficacious as PE in the first-line treatment of this disease.

Topoisomerase I Inhibitors

Camptothecin, a plant alkaloid extract derived from the oriental tree *Comptothecan acuminata,* is a potent inhibitor of DNA topoisomerase I, the nuclear enzyme that relaxes supercoiled DNA by creating single-stranded breaks through which another DNA strand can pass during DNA replication. Due to clinical toxicities observed with the parent compound, three semisynthetic derivatives have been developed which are currently in clinical trials: irinotecan (CPT-11), 9-aminocamptothecin and topotecan. These agents stabilize a covalent DNA-topoisomerase I complex to yield enzyme-linked DNA single-strand breaks. They have been found to be active against a number of tumor models in vitro and in animal xenografts [9–11].

Topotecan. Based upon the clinical activity of topoisomerase II inhibitors such as etoposide in this disease and the preclinical activity of topoisomerase I inhibitors, the Eastern Cooperative Oncology Group (ECOG) conducted a

Table 1. Phase II studies of single-agent topotecan in SCLC

	Number evaluable	Stage	Response rate			Median survival
			CR	PR	overall	
First line						
Schiller [12]	48	ED	0	19 (39.5%)	39.5%	10 months
Second line		relapsed				
'Sensitive'[1]						
Ardizzoni [13]	44		6 (14%)	11 (25%)	39%	6.9 months
Eckardt [14]	52		0	8 (15%)	15%	28 weeks
'Refractory'[2]						
Ardizzoni [13]	42		1 (2%)	2 (5%)	7%	4.7 months
Eckardt[14]	47		0	1 (2%)	2%	21 weeks
Perez-Soler [15]	28		0	3 (11%)	11%	20 weeks

[1] >3 months following first-line chemotherapy regimen.
[2] ≤3 months following first-line chemotherapy regimen.

phase II trial of topotecan in extensive-stage SCLC [12]. Forty-eight patients with previously untreated, extensive-stage SCLC received 2.0 mg/m^2 of topotecan daily for 5 days. The first 13 patients were treated without colony-stimulating-factor support; the next 35 patients received 5 µg/kg of granulocyte-colony-stimulating factor (G-CSF) for 10–14 days starting on day 6. Cycles were repeated every 3 weeks for a maximum of 4 cycles. Patients having a partial response to topotecan after 4 cycles, or stable disease after 2 cycles, or progressive disease at any time, received salvage chemotherapy with PE. Topotecan pharmacokinetics were measured using a 4-point sampling scheme.

Nineteen of the 48 patients had a partial response for an objective response rate of 39.5% (table 1). The median response duration was 4.8 months. After median follow-up of 18.2 months, the overall median survival was 10.0 months; 1-year survival rate was 39% (95% confidence interval (CI): 25.2–53.0%).

Eight of 34 patients (24%) receiving salvage chemotherapy responded. Four of the 17 patients who did not respond to first-line therapy with topotecan responded to PE.

The most common toxicity was hematologic. Ninety-two percent of patients treated without G-CSF developed grade 3 or 4 neutropenia, compared to 29% receiving G-CSF. The incidence of neutropenic fevers was 11 and 8% in the groups with and without G-CSF, respectively, and 1 patient in each group died of infection. Thrombocytopenia and anemia was relatively mild (incidence of grade 3 and 4 thrombocytopenia and anemia of 38 and 27%, respectively). Nonhematologic toxicity was otherwise mild.

There was a correlation between the white blood cell count and absolute neutrophil counts and both the area under the curve (AUC) and maximum concentration (C_{max}) of total topotecan in plasma and topotecan lactone in plasma. However, there was no correlation between the tumor response and the estimated pharmacokinetic parameters.

Other investigators have explored the role of topotecan in previously treated patients with SCLC. The European Organization for Research and Treatment of Cancer (EORTC) studied patients who had either failed first-line treatment ≤ 3 months from chemotherapy discontinuation ('refractory' patients) or had responded to first-line therapy and progressed > 3 months after their chemotherapy was stopped ('sensitive' patients) [13]. One hundred one patients were entered onto the study; 86 were evaluable for response. Patients received 1.5 mg/m^2 for 5 consecutive days every 3 weeks until disease progression or toxicity. The overall response rate was 23%; however, 39% of the 'sensitive' group responded whereas only 7% of the 'refractory' group responded (table 1). The median survival of the 'refractory' patients was 4.7 months while the median survival of the 'sensitive' patients was 6.9 months. Grade 3 and 4 neutropenia occurred in 28 and 47% of the cycles, respectively.

Eckardt et al. [14] conducted a similar trial in 99 patients with SCLC who had failed one prior chemotherapy regimen. Forty-seven patients were 'refractory' to their prior therapy and 52 were 'sensitive', using definitions very similar to those used in the EORTC study. Results were assessed by 'intent-to-treat' analysis and revealed a 2% response rate in 'refractory' patients and a 15% response rate in 'sensitive' patients (table 1). The observed response rates were significantly lower in the US study, perhaps due to differences in an intent-to-treat analysis versus an analysis of evaluable patients only in the EORTC trial, which had a high unevaluability rate. Despite the difference in response rates, however, the median survival was similar; 21 and 28 weeks in the 'refractory' and 'sensitive' groups, respectively. Grade 4 neutropenia with fever was observed in 17% of patients and 5% of cycles.

M.D. Anderson Cancer Center also conducted a phase II trial of topotecan, but restricted their patient population to patients who had failed to respond to etoposide-containing front-line therapy, or who had progressed during or within 3 months of the first dose of front-line or second-line therapy containing etoposide [15]. Three of 28 (11%) evaluable patients achieved a partial remission; 1 of these 3 had failed to respond to front-line cisplatin and etoposide (table 1). The overall median duration of survival was 20 weeks.

Interestingly, topotecan has been reported to have activity on SCLC brain metastases [16]. Topotecan was administered to 16 patients with asymptomatic

Table 2. Phase II studies of irinotecan in SCLC

	Number evaluable	Stage	Response rate			Median survival
			CR	PR	overall	
First line						
(plus cisplatin)						
Fujiwara [18]	32					
	18	LD	4	10	78%	–
	14	ED	3	8	69%	–
Second line		relapsed				
'Sensitive'[1]						
Masuda [17]	15		0	7	47%	27 weeks

[1] >3 months following first-line chemotherapy regimen.

SCLC brain metastases. Four of 16 had a complete response and 6 of 16 had a partial response, with a CNS response rate of 63%. Responses were observed both systemically, as well as in the brain.

Irinotecan. Irinotecan is another topoisomerase I inhibitor which has been shown to have activity in SCLC. Masuda et al. [17] conducted a phase II trial of irinotecan as a second-line therapy in patients with refractory or relapsed SCLC. Sixteen patients received 100 mg/m^2 as a 90-min infusion every week; 15 patients were evaluable for toxicity, response, and survival. Fourteen patients had relapsed >3months after completion of their first-line chemotherapy. Seven patients responded (47%; 95% CI: 21.4–71.9%) with a median duration of response of 58 days (table 2). Median survival was 27 weeks. Thirty-three percent had grade 3 or 4 leukopenia, and 13% had grade 3 or 4 pulmonary toxicity as evidenced by dyspnea on exertion, a diffuse reticulo-nodular pattern on chest x-ray, and high fever. One patient had grade 3 diarrhea.

Based upon these results, a phase II study of irinotecan and cisplatin was conducted in previously untreated patients with SCLC [18]. Patients were to receive 80 mg/m^2 irinotecan as a 90-min infusion on days 1, 8 and 15 in combination with 60 mg/m^2 of cisplatin; however, the dose of irinotecan was reduced to 60 mg/m^2 after 10 patients because of severe hematologic toxicity, diarrhea, or hepatic toxicity. Of the 32 patients eligible for response, 25 responded (14 of 18 limited-stage patients and 11 of 14 extensive-stage patients) for an overall response rate of 78% (table 2).

These data suggest that the topoisomerase inhibitors have activity in SCLC which warrants further investigation. Additional studies are needed to determine the optimal method of combining and sequencing these agents with other cytotoxic drugs for SCLC, particularly topoisomerase II inhibitors such as etoposide.

Taxanes

The taxanes, paclitaxel (Taxol®) and docetaxel (Taxotere®), have generated considerable enthusiasm because of their broad spectrum activity against a variety of tumor types. Derived from the western yew tree *Taxus brevifolia* and the European yew tree *Taxus baccata,* respectively, these agents enhance tubulin assembly into microtubules and inhibit depolymerization of micro-tubules. They have preclinical activity against a wide variety of tumor types, and have clinical activity against breast cancer, non-small cell lung cancer (NSCLC), and ovarian carcinoma.

Paclitaxel. Paclitaxel has been evaluated as a single agent in previously untreated patients with extensive-stage SCLC in several phase II studies. The ECOG treated 36 patients with 250 mg/m² of paclitaxel over 24 h every 3 weeks. Nonresponders or partial responders who received the maximum number of 4 cycles of paclitaxel were given salvage therapy with etoposide and cisplatin [19]. Of 32 patients evaluable for response, 11 (34%) had a partial response (table 3). In 3 of 6 patients categorized as having stable disease, there was > 50% tumor shrinkage, but no 4-week measurements were made. The median duration of survival was 43 weeks. Nineteen patients (56%) experienced grade 4 neutropenia; 1 patient each experienced grade 4 cardiac and hypersensitivity reactions.

Kirschling et al. [20] conducted a study of 250 mg/m² of paclitaxel over a 24-hour infusion every 21 days to previously untreated patients with exten-sive-stage SCLC, but also administered prophylactic G-CSF. In a preliminary report presented in abstract form, no complete responses and 15 partial re-sponses were observed in 37 evaluable patients (41%) with a median survival of 29 weeks (table 3). Grade 4 leukopenia was observed in 19% of paclitaxel courses.

Given the activity of single-agent paclitaxel, phase I and II trials are ongoing to determine the activity of paclitaxel in combination with other cytotoxic drugs, particularly etoposide and cisplatin. Bunn and Kelly [21] conducted a phase I study of cisplatin, etoposide, and paclitaxel in patients with advanced, previously untreated SCLC. Paclitaxel was administered as a 3-hour infusion prior to PE. The starting doses of cisplatin were 80 mg/m² day 1, 50 mg/m² of etoposide i.v. day 1 and 100 mg/m² orally days 2 and 3; and 135 mg/m² paclitaxel. Of 13 patients described in the preliminary report,

Table 3. Phase II studies of taxanes in SCLC

	Stage	Paclitaxel	Other drugs	n	CR	PR	Median survival
Paclitaxel – previously untreated							
Ettinger [19]	ED	250 mg/m² over 24 h every 3 weeks		32	0	11 (34%)	43 weeks
Kirschling [20]	ED	250 mg/m² over 24 h every 3 weeks	G-CSF	37	0	15 (41%)	29 weeks
Hainsworth [22, 23]	ED +LD	135 mg/m² over 1 h every 3 weeks	Carboplatin – AUC 5 day 1 Etoposide – 100 mg alt. with 50 mg p.o. days 1–10 *or* 25 mg i.v. days 1–5 and 8–12	38	10 (26%)	19 (50%)	ED – 7 months LD – 17 months
Docetaxel – previously treated							
Smyth [26]	relapsed	100 mg/m² over 1 h every 3 weeks		28	0	7 (25%)	–

9 were evaluable for response. Five complete responses and 4 partial responses were observed; no grade 4 toxicities with the exception of neutropenia were reported.

Hainsworth et al. [22] combined paclitaxel with carboplatin and etoposide as a 1-hour infusion in two phase II studies. The first used paclitaxel at a dose of 135 mg/m² day 1, carboplatin (GFR + 25) times 5, day 1, and etoposide 25 mg/m² i.v. days 1–5 and 8–12; concurrent radiation therapy was administered to patients with limited-stage disease [21]. In the second study, patients received carboplatin at an AUC of 5 and oral etoposide, 100 mg alternating with 50 mg, on days 1–10 in addition to the paclitaxel [23]. Of the 38 patients entered on the two studies, 29 responded (76%), including 26% of patients who had complete responses (table 3). Responses were observed in 93% of patients with limited-stage disease, and 15 of 23 patients with extensive-stage disease. Median survival was 7 months for patients with extensive-stage disease, and 17 months for limited-disease patients. Eleven episodes of grade 3 or 4 neutropenia were observed (8% of courses), although 15 patients required hospitalization for leukopenia.

Paclitaxel has also been studied as a 4-day continuous infusion followed by cisplatin in a phase I study involving patients with both NSCLC and SCLC [24].

Four of 6 previously untreated SCLC patients had partial responses. In a phase I study involving ifosfamide, carboplatin, etoposide and 24-hour infusion of paclitaxel, 5 of 6 patients with SCLC responded, including 2 complete responses [25]. Twenty-six percent of the SCLC and advanced NSCLC patients had fever and neutropenia in the latter study, and 63 and 49% had grade 4 neutropenia and thrombocytopenia, respectively. In both studies, additional responses were observed in NSCLC; however, data presented were not mature.

Docetaxel. The EORTC explored the activity of docetaxel in patients with previously treated SCLC. Thirty-four patients received 100 mg/m^2 of docetaxel over 1 h every 21 days [26]. Although no information was given regarding the response to first-line therapy, 7 partial responses were reported to docetaxel in 28 evaluable patients (25%). Six patients were excluded from response analysis for having completed less than two courses (3 died of sepsis, 1 patient developed a severe hypersensitivity reaction, and there were 2 early deaths). Two additional patients had >50% shrinkage of disease which was not sustained for the minimally acceptable time. Median duration of response was 4.7 months; no survival data was given. Twenty-two patients developed grade 4 neutropenia.

These data suggest taxanes also have single-agent activity in SCLC which warrants further investigation. Additional studies are ongoing to determine the role of these drugs when combined with other cytotoxic agents in the management of SCLC.

Gemcitabine

Gemcitabine, difluorodeoxycytidine, is a pyrimidine antimetabolite which is an analog of deoxycytidine. In addition to clinical activity in pancreatic and NSCLC [27], preclinical antitumor effects have also been observed in SCLC xenografts [28]. To further determine the activity of this drug in SCLC, the National Cancer Institute of Canada (NCI-C) conducted a phase II trial of gemcitabine in 29 previously untreated patients with extensive-stage SCLC [29]. The first 17 patients received 1,000 mg/m^2/week and the remaining 12 patients 1,250 mg/m^2/week. Patients who failed to respond after 2 cycles of therapy were offered standard therapy. One complete and 6 partial responses were observed in 26 patients, for an overall response rate of 27%. Median response duration was 12.5 weeks and median survival was 12 months. Grade 3 or 4 neutropenia was observed in only 13 of 72 cycles (18%). Fatigue, anorexia and nausea were observed in 27.5, 27.5 and 72% of patients, respectively. Of the 22 non-responding patients, 19 received second-line therapy and 12 (63%) responded. These results suggest that gemcitabine may have activity in this disease with only modest toxicity. The high second-line response rate suggests that clinical drug resistance did not develop. Additional studies are ongoing which combine gemcitabine with other cytotoxic agents as a first-line regimen, and exploring the role of gemcitabine as a single agent for relapsing disease.

Chronic Oral Etoposide

Etoposide, or VP-16, is a semisynthetic derivative of podophyllotoxin, a topoisomerase II inhibitor which results in single-strand breaks in DNA. It is one of the most active single agents in SCLC, with an overall response rate of approximately 45% [30]. When administered in a 'standard' regimen over 3 days with cisplatin, it induces responses in about 60–80% of extensive-stage SCLC patients and over 90% of limited-disease patients [3, 31].

Earlier schedule-specific in vitro [32, 33] studies suggested that prolonged schedules of administration are more efficacious than single, high-dose administration [34–36]. An early trial of etoposide scheduling was conducted in 60 patients with SCLC, 45 of whom were untreated [34]. Patients were randomized to 3 schedules, in which they received 250 mg/m² i.v. weekly, 500 mg/m² orally over 3 days weekly, or 850 mg/m² orally over 5 days every 3 weeks. The response rates were 20, 65 and 42%, respectively, suggesting that more prolonged schedules of etoposide were superior. These results were confirmed in two prospective trials in previously untreated SCLC. Slevin et al. [35] randomized 39 patients with extensive disease to receive 500 mg/m² as a continuous infusion over 24 h or to receive 5 consecutive daily 2-hour infusions each of 100 mg/m². The response rates were dramatically different, at 10 and 89%, respectively. Abratt et al. [36] randomized patients with limited-stage disease to receive one of two combination regimens that differed only in the scheduling of etoposide. Patients in one arm received 60 mg/m² i.v. on day 1 and 120 mg/m² orally on days 2–5 of each cycle, while patients in the other arm received 300 mg/m² of etoposide i.v. on day 1. The complete (53 vs. 26%) and overall (75 vs. 52%) response rates were significantly higher in the oral etoposide arm.

Several phase II studies have been conducted to explore the tolerability and efficacy of prolonged administration of oral etoposide [37]. The results of these studies suggest that oral etoposide is well tolerated and has significant therapeutic activity in patients with relapsed or refractory SCLC, including those who have received prior intravenous etoposide [38, 39]. For example, a phase II trial of etoposide was conducted in patients with relapsed or refractory SCLC. Twenty-two patients (18 of whom had received prior i.v. etoposide) received 50 mg/m²/day for 21 days. Ten of 22 patients achieved a complete or partial response [39].

Oral Etoposide plus Intravenous Cisplatin. These promising results have generated interest in combining chronic oral etoposide with cisplatin as first-line treatment for SCLC. The ECOG conducted a phase II trial of oral etoposide and cisplatin in newly diagnosed, untreated patients with extensive-stage small cell carcinoma of the lung [40]. Thirty-five patients received 100 mg/m^2 of cisplatin i.v. on day 1 and 50 mg/m^2 of etoposide orally for 21 consecutive days. Cycles were repeated every 28 days. The most common toxicity observed was myelosuppression. Sixty-seven percent of patients had grade 3 or 4 leukopenia and 34% had grade 3 or 4 thrombocytopenia during cycle 1. Twenty-eight patients had dose reductions or delays, or had their treatment stopped early due to myelosuppression. Four patients died of neutropenic sepsis. Of 26 evaluable patients, 4 had a complete response (15%) and 17 had a partial response (65%). The median survival for the group as a whole was 8.5 months.

The Cancer and Leukemia Group B conducted a randomized phase III study of 21-day oral etoposide versus 3 days i.v. etoposide in combination with i.v. cisplatin in extensive-stage SCLC [41]. One hundred fifty-six patients were randomized to receive etoposide (130 mg/m^2) i.v. daily for 3 days and 25 mg/m^2 of cisplatin i.v. for 3 days every 21 days for 8 courses, and 150 patients were randomized to receive 50 mg/m^2/day orally for 21 days and 33 mg/m^2 of cisplatin i.v. for 3 days every 28 days for 6 courses. Median survival was 9.5 and 9.9 months, respectively. Overall response rates were 57 and 61%, respectively. Grade 4 neutropenia was observed in 85% of patients randomized to i.v. etoposide, and 83% of patients randomized to oral etoposide. Four percent and 10% of patients on the i.v. and oral etoposide arms, respectively, died of neutropenic fevers (not statistically significant). These two studies suggest that oral etoposide, when combined with i.v. cisplatin, is associated with significant myelosuppression, and offers no therapeutic advantage to other commonly administered chemotherapeutic regimens for SCLC.

Oral Etoposide in the Elderly or Poor Performance Status Patients. In addition to evaluating the use of chronic oral etoposide in combination with cisplatin as a first-line agent, oral etoposide has also been evaluated as a single first-line agent for the elderly, or patients with poor performance status. In elderly, untreated patients with SCLC, phase II trials of 800 mg/m^2 of etoposide over 5 days orally resulted in a 79% response rate and was very well tolerated, with minimal myelosuppression [42].

The Medical Research Council (MRC) of the United Kingdom compared the efficacy of a 10-day regimen of oral etoposide with multidrug chemotherapy in palliating symptoms, impact on quality of life, response, and survival in patients with poor performance status [43]. Patients were required to have no prior therapy and performance status of 2–4. The original objectives of the protocol were to randomize 450 patients to recieve either 4 cycles of 50 mg of etoposide orally twice daily for 10 days, or a standard regimen consisting either of etoposide plus vincristine, or cyclophosphamide, doxorubicin, and vincristine. The etoposide in the etoposide-containing arm consisted of 120 mg/m^2 i.v. day 1, with either 240 mg/m^2 orally or 120 mg/m^2 i.v. days 2 and 3. However, patient accrual was stopped early at 339 patients when an interim analysis showed that the palliative effects of treatment were similar between the two groups (41 vs. 46%), but there was enhanced hematological toxicity (29% grade 2 or worse hematological toxicity in the etoposide-treated patients versus 21% in the control group). Furthermore, significantly worse response rate (45 vs. 51%) and survival (130 vs. 183 days) was observed in the oral etoposide group compared to the control groups, respectively.

A second study by the London Lung Cancer Group compared 100 mg of oral etoposide twice daily for 5 days with cisplatin/etoposide (60 mg/m^2 of cisplating day 1 with 120 mg/m^2 of etoposide i.v. day 1 and 100 mg twice daily orally days 2 and 3) alternating with cyclophosphamide, doxorubicin, and vincristine [44]. To be eligible, patients had to be either over 75 years of age, or had to have an ECOG performance status of 2–3 or an alkaline phosphatase > 1.5 times the upper limits of normal. One hundred fifty-five patients were randomized. The overall response rate was 61% in the combination chemotherapy arm with 39% in the oral etoposide arm. Progression-free survival was in favor of the combination chemotherapy arm (170 vs. 111 days); median overall survival showed a trend in favor of the combination chemotherapy arm (189 vs. 146 days) which was not statistically superior. Quality-of-life analysis showed improvement in the i.v. arm, with the exception of nausea and vomiting. Doses reductions and delays were more common in the i.v. arm secondary to hematological or renal toxicity, although grade 3 or 4 toxicities were rare in either arm, with the exception of nausea and vomiting. Similar to the MRC study, the accrual to the protocol was stopped early following an interim analysis.

Taken together, these studies show that oral etoposide is not as effective as combination chemotherapy regimens at either palliating symptoms or prolonging survival in the elderly, or patients with a poor performance status.

Ifosfamide

Ifosfamide, an analog of the alkylating agent cyclophosphamide, is one of the most active agents in the treatment of SCLC. As a single agent, the drug produces a > 50% response rate [45]. A number of phase II studies have reported response rates of 74–100% in limited- and/or extensive-stage disease when combined with cisplatin and etoposide (VIP) or carboplatin and etoposide (ICE) [46–52].

The Hoosier Oncology Group (HOG) recently completed a randomized phase III study of PE with and without ifosfamide in extensive-stage SCLC [53]. Patients were previously untreated, and allowed to have a Karnofsky performance status of ≥ 50 and CNS metastases. Patients were randomized to receive 20 mg/m^2/day of cisplatin and 100 mg/m^2/day of etoposide on days 1–4 (VP), or 20 mg/m^2/day cisplatin, 1.2 g/m^2/day ifosfamide, and 75 mg/m^2/day etoposide days 1–4, with mesna cytoprotection (VIP), every 3 weeks for 4 cycles. One hundred seventy-one patients were randomized. Fifty-five of 82 evaluable patients treated with VP (67%) and 59 of 81 patients treated with VIP (73%) responded ($p > 0.05$). The median survival times for the VP and VIP arms were 7.3 and 9.0 months, respectively ($p = 0.045$) with 2-year survival rates of 5 versus 13%, respectively. Myelosuppression was significant in both arms, with 20 of 82 (24%) patients in VP arm experiencing grade 4 neutropenia, and 34 of 80 (42%) patients in the VIP arm experiencing grade 4 neutropenia. Eight patients in the VP arm and 9 patients in VIP arm were hospitalized for febrile neutropenia; there were 11 early deaths (6 treated with VP and 5 with VIP). The authors conclude that the addition of ifosfamide to standard PE therapy was associated with a modest improvement in survival, and a moderate increase in toxicity.

New trials are evaluating the role of ifosfamide-containing regimens with paclitaxel [25] or oral etoposide. The HOG conducted a phase II study of daily oral etoposide (37.5 mg/m^2 × 21 days) plus ifosfamide (1.2 g/m^2/day for 4 days) and cisplatin (20 mg/m^2/day for 4 days) for previously treated patients with recurrent disease. Courses were repeated every 4 weeks [54]. Thirty-six of 42 patients had received prior PE although patients were excluded if they had received prior ifosfamide, daily oral etoposide, or had progressed with 4 weeks of prior therapy. The overall response rate was 55%; the median progression-free survival time was 20 weeks, and the overall median survival duration was 29 weeks. However, myelosuppression was significant. After the first 22 patients were entered, the protocol was changed so that the subsequent 20 patients received oral etoposide for 14 days. Fourteen of 22 patients (64%) treated with 21 days of oral etoposide developed grade 4 neutropenia, and 16 of 20 patients (80%) treated with 14 days of oral etoposide developed grade 4 neutropenia. There were 6 treatment-related deaths, including 4 deaths as a result of sepsis; 29% were hospitalized for neutropenic fevers.

Conclusion

Hopefully, the continued development of new drugs discussed here will make an impact in the lives of patients with SCLC. To further elucidate their role in the treatment of this disease, continued efforts need to be made at enrolling patients in clinical trials. Additional efforts also need to be made at understanding and exploring the biology of this disease, so that our ultimate efforts in controlling this cancer will combine cytotoxic therapies and interventions at the molecular or biological level.

References

1   Evans WK, Feld R, Murray N, Willan A, Coy P, Osoba D, Shepherd FA, Clark DA, Levitt M, MacDonald A: Superiority of alternating non-cross-resistant chemotherapy in extensive small cell lung cancer. A multicenter, randomized clinical trial by the National Cancer Institute of Canada. Ann Intern Med 1987;107:451–458.
2   Fukuoka M, Furuse K, Saijo N, Nishiwaki Y, Ikegami H, Tamura T, Shimoyama M, Suemasu K: Randomized trial of cyclophosphamide, doxorubicin, and vincristine versus cisplatin and etoposide versus alternation of these regimens in small-cell lung cancer. J Natl Cancer Inst 1991;83: 855–861.
3   Roth B, Johnson D, Einhorn L, Schacter L, Cherng N, Cohen H, Crawford J, Randolph J, Goodlow J, Broun G, Omura G, Greco F: Randomized study of cyclophosphamide, doxorubicin, and vincristine versus etoposide and cisplatin versus alternation of these two regimens in extensive small-cell lung cancer: A phase III trial of the Southeastern Cancer Study Group. J Clin Oncol 1992;10: 282–291.
4   Feld R, Evans WK, Coy P, Hodson I, MacDonald AS, Osoba D, Payne D, Shelley W, Pater JL: Canadian multicenter randomized trial comparing sequential and alternating administration of two non-cross-resistant chemotherapy combinations in patients with limited small-cell carcinoma of the lung. J Clin Oncol 1987;5:1401–1409.
5   Ettinger D, Finkelstein D, Abeloff M, Bonomi P: Justification for evaluating new anti-cancer drugs in selected untreated patients with a chemotherapy-sensitive advanced cancer: An ECOG randomized study. Proc Am Soc Clin Oncol 1990;9:224.
6   Ettinger D, Finkelstein D, Abeloff M, Ruchdeschel J, Aisner S, Eggleston J: A randomized comparison of standard chemotherapy versus alternating chemotherapy and maintenance versus no maintenance therapy for extensive-stage small-cell lung cancer: A phase II study of the Eastern Cooperative Oncology Group. J Clin Oncol 1990;8:230–240.
7   Moore T, Korn E: Phase II trial design considerations for small-cell lung cancer. J Natl Cancer Inst 1992;84:150–154.
8   Skarlos D, Samantas E, Kosmidis P, Fountzilas G, Angelidou M, Palamidas P, Mylonakis N, Provata A, Papadakis E, Klouvas G, Theocharis D, Panousaki E, Boleti E: Randomized comparison of etoposide-cisplatin vs. etoposide-carboplatin and irradiation in small-cell lung cancer. Ann Oncol 1994;5:601–607.
9   Liu L: DNA topoisomerase poisons as antitumor drugs. Annu Rev Biochem 1989;58:351–375.
10   Chabner B: Camptothecins. J Clin Oncol 1992;10:3–4.
11   Rowinsky E, Grochow L, Hendricks C, Ettinger D, Forastiere A, Hurowitz L, McGuire W, Sartorius S, Lubejko B, Kaufmann S, Donehower R: Phase I and pharmacologic study of topotecan: A novel topoisomerase I inhibitor. J Clin Oncol 1992;10:647–656.
12   Schiller J, Kim K, Hutson P, DeVore R, Glick J, Stewart J, Johnson D: Phase II study of topotecan in patients with extensive-stage small cell carcinoma of the lung; An Eastern Cooperative Oncology Group Trial (E1592). J Clin Oncol 1996;14:2345–2352.
13   Ardizzoni A, Hansen H, Wanders J, Dombernowsky P, Gamucci T, Kaplan S, Postmus P, Hudson I, Giaccone G, Verweij J: Topotecan, a new active agent in the second-line treatment of 'refractory' and 'sensitive' small-cell lung cancer. Ann Oncol 1996;7:106.
14   Eckardt J, Gralla R, Palmer MC, Gandara D, Laplante J, Sandler A, Fields SZ, Fitts D, Broom C: Topotecan as second-line therapy in patients with small cell lung cancer: A phase II study. Ann Oncol 1996;7:107.

15 Perez-Soler R, Glisson BS, Lee JS, Fossella FV, Murphy WK, Shin DM, Hong WK: Treatment of patients with small-cell lung cancer refractory to etoposide and cisplatin with the topoisomerase I poison topotecan. J Clin Oncol 1996;14:2785–2790.

16 Managold C, Pawel JV, Scheithauer W, Lan J, Schaefer B, Pastovic R, Staab HJ: Response of SCLC brain metastases on topotecan (SK&F 104864) therapy. Ann Oncol 1996;7:106.

17 Masuda N, Fukuoka M, Kusunoki Y, Matsui K, Takifuji N, Kudoh S, Negoro S, Nishioka M, Nakagawa K, Takada M: CPT-11: A new derivative of camptothecin for the treatment of refractory or relapsed small-cell lung cancer. J Clin Oncol 1992;10:1225–1229.

18 Fujiwara Y, Yamakido M, Fukuoka M, Kudoh S, Furuse K, Ikegami H, Ariyoshi Y: Phase II study of irinotecan and cisplatin in patients with small cell lung cancer. Proc Am Soc Clin Oncol 1994;13:335.

19 Ettinger D, Finkelstein D, Sarma R, Johnson D: Phase II study of paclitaxel in patients with extensive-disease small-cell lung cancer: An Eastern Cooperative Oncology Group study. J Clin Oncol 1995;13:1430–1435.

20 Kirschling RJ, Jung SH, Jett JR, NCCT Group: A phase II trial of taxol and GCSF in previously untreated patients with extensive stage small cell lung cancer. Proc Am Soc Clin Oncol 1994;13:326.

21 Bunn PJ, Kelly K: A phase I study of carboplatin and paclitaxel in small-cell lung cancer: A University of Colorado Cancer Center study. Semin Oncol 1995;22:54–58.

22 Hainsworth JD, Erland J, Peters M, Greco FA: Treatment of small cell lung cancer with taxol (1-hour infusion), carboplatin, and etoposide: Phase II study of an outpatient regimen. Proc Am Soc Clin Oncol 1994;13:135.

23 Hainsworth JD, Stroup SL, Greco FA: Paclitaxel, carboplatin, and extended schedule etoposide in the treatment of small cell lung carcinoma. Cancer 1996;77:2458–2463.

24 Georgiadis MS, Brown JE, Schuler BS, Frame JN: Phase I study of a four-day continuous infusion of paclitaxel followed by cisplatin in patients with advanced lung cancer. Proc Am Soc Clin Oncol 1995;14:A1072.

25 Strauss GM, Lynch TJ, Elias AD, Jacobs C, Kwiatkowski DJ, Shulman LN, Carey RW, Grossbard ML, Jauss S, Sugarbaker DJ: A phase I study of ifosfamide/carboplatin/etoposide/paclitaxel in advanced lung cancer. Semin Oncol 1995;22:70–74.

26 Smyth JF, Smith IE, Sessa C, Schoffski P, Wanders J, Franklin H, Kaye SF: Activity of docetaxel (taxotere) in small cell lung cancer. Eur J Cancer 1994;30A:1058–1060.

27 Kaye SB: Gemcitabine: Current status of phase I and II trials. J Clin Oncol 1994;12:1527–1531.

28 Kristjansen PEG, Quistorff B, Spang-Thomsen M, Hansen HH: Intratumoral pharmacokinetic analysis by 19f-magnetic resonance spectroscopy and cytostatic in vivo activity of gemcitabine (dfdc) in two small cell lung cancer xenografts. Ann Oncol 1993;4:157–160.

29 Cormier Y, Eisenhauer E, Muldal A, Gregg R, Ayoub J, Goss G, Stewart D, Tarasoff P, Wong D: Gemcitabine is an active new agent in previously untreated extensive small cell lung cancer: A study of the National Cancer Institute of Canada Clinical Trials Group. Ann Oncol 1994;5:283–285.

30 Comis RL: Chemotherapy of small cell lung cancer. Principles Practise Oncol Updates 1987;1:1–18.

31 Boni C, Cocconi G, Bisagni G, Ceci G, Peracchia G: Cisplatin and etoposide (VP-16) as a single regimen for small cell lung cancer: A phase II trial. Cancer 1989;63:638–642.

32 Dombernowsky P, Nissen N: Schedule dependency of the antileukemic activity of the podophyllo-toxin-derivative VP-16-213 (NSC-141540) in L1210 leukemia. Acta Pathol Microbiol Scand 1973; 81:715–724.

33 Fleming R, Miller A, Stewart C: Etoposide: An update. Clin Pharm 1989;8:274–293.

34 Cavalli F, Sonntag R, Senn H, Brunner K: VP-16-213 monotherapy for remission induction of small cell lung cancer: A randomized trial using three dosage schedules. Cancer Treat Rep 1978; 62:473–475.

35 Slevin M, Clark P, Joel S, Malik S, Osvorne R, Gregory W, Lowe D, Reznek R, Wrigley P: A randomized trial to evaluate the effect of schedule on the activity of etoposide in small-cell lung cancer. J Clin Oncol 1989;7:1333–1340.

36 Abratt R, Willcox P, de Groot M, Goodman H, Jansen E, Mame Salton D: Prospective study of etoposide scheduling in combination chemotherapy for limited disease small cell lung carcinoma. Eur J Cancer 1991;27:28–30.

37   Hainsworth J, Johnson D, Frazier S, Greco F: Chronic daily administration of oral etoposide – A phase I trial. J Clin Oncol 1989;7:396–401.
38   Einhorn L, Pennington K, McClean J: Phase II trial of daily oral VP-16 in refractory small cell lung cancer: A Hoosier Oncology Group study. Semin Oncol 1990;17:32–35.
39   Johnson D, Greco F, Strupp J, Hande K, Hainsworth J: Prolonged administration of oral etoposide in patients with relapsed or refractory small-cell lung cancer: A phase II trial. J Clin Oncol 1990; 8:1613–1617.
40   Schiller J, Johnson D: Phase II trial of oral etoposide plus cisplatin in extensive stage small cell carcinoma of the lung: An Eastern Cooperative Oncology Group study. Eur J Cancer 1994;30A: 158–161.
41   Miller AA, Herndon JEI, Hollis DR, Ellerton J, Langleben A, Richards FI, Green MR: Schedule dependency of 21-day oral versus 3-day intravenous etoposide in combination with intravenous cisplatin in extensive-stage small-cell lung cancer: A randomized phase III study of the Cancer and Leukemia Group B. J Clin Oncol 1995;13:1871–1879.
42   Carney D, Grogan L, Smit E, Harford P, Berendsen H, Postmus P: Single-agent oral etoposide for elderly small cell lung cancer patients. Semin Oncol 1990;17:49–53.
43   Medical Research Council of the United Kingdom Party: Comparison of oral etoposide and standard intravenous multidrug chemotherapy for small-cell lung cancer: A stopped multicentre randomised trial. Lancet 1996;348:563–566.
44   Harper P, Underhill C, Ruiz de Elvira MC, Rudd R, Souhami R, Spiro S, Trask C, Partridge M, Tobias J, Ledermann J, Eraut D, James L: A randomised study of oral etoposide versus combination chemotherapy in poor prognosis small cell lung cancer. Proc Am Soc Clin Oncol 1996;15(abstr 2019).
45   Drings P: Ifosfamide in the treatment of bronchial carcinoma. Contrib Oncol 1987;26:294–318.
46   Evans W, Stewart D, Shepherd F: VP-16, ifosfamide and cisplatin (VIP) for extensive small cell lung cancer. Eur J Cancer 1994;30A:299–303.
47   Loehrer PJ, Rynard S, Ansari R: Etoposide, ifosfamide, and cisplatin in extensive small cell lung cancer. Cancer 1992;69:669–673.
48   Munoz MA, Arrivi A, Guaraz R: VP-16, ifosfamide and cisplatin (VIP) as treatment of small cell lung cancer: A preliminary report. Fourth European Conference on Clinical Oncology and Cancer Nursing, Madrid 1987.
49   Carney DN, Grogan L: Phase II study of VP-16, ifosfamide and cisplatin (VIP) in small cell and nonsmall cell lung cancer. Br J Cancer 1990;62:523.
50   Thatcher N, Lind M, Stout R, Payne C, Carroll KB, Campbell C, Moussalli H: Carboplatin, ifosfamide, and etoposide with mid-course vincristine and thoracic radiotherapy for 'limited' stage small cell carcinoma of the bronchus. Br J Cancer 1989;60:90–101.
51   Wolff A, Ettinger D, Neuberg D, Comis R, Ruckdeschel J, Bonomi P, Johnson D: A phase II study of ifosfamide, carboplatin, and oral etoposide chemotherapy for extensive-disease small cell lung cancer: An ECOG pilot study. Proc Am Soc Clin Oncol 1994;13:357.
52   Smith IE, Perren TJ, Ashley SA: Carboplatin, etoposide, and ifosfamide as intensive chemotherapy for small-cell lung cancer. J Clin Oncol 1990;8:899–905.
53   Loehrer PJ, Sr, Ansari R, Gonin R, Monaco F, Fisher W, Sandler A, Einhorn LH: Cisplatin plus etoposide with and without ifosfamide in extensive small-cell lung cancer: A Hoosier Oncology Group study. J Clin Oncol 1995;13:2594–2599.
54   Faylona EA, Loehrer PJ, Ansari R, Sandler AB, Gonin R, Einhorn LH: Phase II study of daily oral etoposide plus ifosfamide plus cisplatin for previously treated recurrent small-cell lung cancer: A Hoosier Oncology Group Trial. J Clin Oncol 1995;13:1209–1214.

Joan H. Schiller, MD, K4/666 Clinical Science Center,
600 Highland Avenue, University of Wisconsin, Madison, WI 53792 (USA)
Tel. (608) 263-8600, Fax (608) 263-8613

Schiller JH (ed): Updates in Advances in Lung Cancer. Prog Respir Res.
Basel, Karger, 1997, vol 29, pp 150–172

Chapter 10
··········••••••••••••••••••
Palliative Medicine in Lung Cancer

James F. Cleary

Medical Oncology, University of Wisconsin Medical School, and
Palliative Medicine Service, University Hospital and Clinics, Madison, Wisc., USA

Despite advances in the treatment of lung cancer, the disease still represents a major cause of death throughout the world. In the USA, there are 170,000 new cases diagnosed each year and it represents one third of cancer-associated deaths with 157,400 people dying of the disease in 1995 [1]. The average 5-year survival for all stages and types of lung cancer is 11–14%. The palliation of symptoms of lung cancer patients should be of great importance throughout all stages of the disease and not just in those who are close to death. In this chapter, I will address the range and incidence of symptoms in patients with lung cancer, review the assessment of these symptoms and review treatment options of them.

Symptoms Associated with Lung Cancer

It is essential in considering the symptoms that need palliation to review potential causes of these symptoms in relation to the disease (table 1). The predominant symptoms associated with lung cancer include pain, dyspnea, cough, confusion and fatigue. The incidence of these symptoms has been documented in different institutions. In a review of 289 patients with non-small cell lung cancer (80% of whom were men), cough (79%) and breathlessness (75%) exceeded pain (50%) as the predominant symptom (table 2). This may have been related to the better performance status and longer survival in these patients. The Yorkshire Regional Cancer Group (table 3) had physicians assess the incidence of symptoms in non-small lung cancer patients at the time of presentation. Of particular note, these symptoms were reported

Table 1. Causes of symptoms of lung cancer

Pain
Local lung cancer
Metastatic
     Bone
     Liver
     Head
     Neuropathic pain
Treatment-related
Paraneoplastic – hypertrophic pulmonary arthropathy
Nonneoplastic – other pain syndromes
  (usually subacute-chronic)

Dyspnea and Cough
Tumor
Pleural effusion
Superior vena cava obstruction
Pulmonary embolus (tumor-associated) often acute

Confusion
CNS metastases
Hypoxia
Hypercalcemia
SIADH
Opioids

Fever
Tumor
Infection

---

by patients presenting early in their disease to 4 thoracic physicians and 2 radiotherapists.

Portenoy et al. [4] specifically addressed the incidence of pain in 145 ambulatory lung cancer patients, many of whom were receiving antitumor treatment at Memorial Sloan Kettering Cancer Center. Fifty-seven (39%) of these patients had experienced persistent or frequent cancer-related pain during the previous 2 weeks and 17 (11.7%) had reported pain that was unrelated to lung cancer. Forty-seven of the 145 patients, most of whom had locoregional or metastatic disease, were ambulatory and functioned independently, agreed to further study. One third of the 47 experienced more than one discrete site of cancer associated pain and 10 had three or more sites of discrete pain. Using a 0–100 pain scale, the mean pain scores in the previous 2 weeks were rated at 43.5 ± 24.5 for pain in general, 61.0 ± 27.5 for pain at worst. Ninety

Table 2. Numbers (and percentages) of the 289 patients
with NSCLC reporting particular symptoms at presentation
to the Yorkshire Regional Cancer Organization [3]

	Patients		Symptoms			
			severe		moderate	
	n	%	n	%	n	%
Cough	228	79	12	4	101	35
Hemoptysis	101	35	4	1	30	10
Breathlessness	216	75	23	8	95	33
Chest pain	107	37	9	3	40	14
Other pain	37	13	6	2	14	5
Hoarseness	32	11	6	2	6	2
Anorexia	130	45	10	3	47	16
Malaise	136	47	6	2	43	15
Dysphagia	20	7	0		6	2
Abdominal pain	15	5	0		9	3
Headache	22	8	0		4	1
Others	49	17	4	1	38	13

percent reported that pain was present more than three quarters of the time
and more than 75% of patients reported that worst pain was present for at
least one quarter of the time. Approximately one half of patients reported
moderate or greater interference by pain in activity (46.8%) work (55.3%),
sleep (55.3%), mood (55.3%) and enjoyment of life (57.4%). There was lesser
interference with walking (31.9%) and social relations (40.4%).

Symptom Assessment

Much of the work on symptom assessment has been directed primarily
at pain. Many of the scales use numeric rating indexes or visual analogue
scales. While these are suitable for patients who are able to converse, other
tools must be considered in patients who have difficulty with communication.
This may include those with CNS disease, particularly the elderly.

Pain Assessment
Inadequate pain assessment and poor physician–patient communication
about pain are major barriers to good pain care [5]. Proper pain management

Table 3. Mean symptom assessment scores (0–100) and standard deviations (SD) for 305 patients with unresectable lung cancer both before and during treatment [2]

	Before treatment		During treatment	
	mean score	SD	mean score	SD
Fatigue	39.4	25.4	45.6	25.8
Nausea and vomiting	6.7	15.5	19.6	25.1
Pain	29.3	30.8	22.3	25.7
Dyspnea	41	28.7	37.5	28.6
Sleep disturbance	31.9	33.1	30.9	32.5
Appetite loss	26.9	35.1	32.6	37
Constipation	20.1	31	23.7	32.6
Diarrhea	4.2	14	4.6	14.4

requires a clear understanding of the characteristics of the pain and its physical basis. The changing expression of pain associated with lung cancer demands *repeated assessment*, as new pain sites can emerge rapidly. The essentials of cancer pain assessment are similar to those taught for disease assessment in the early years of professional education, principles used extensively for diagnostic purposes in patients with ischemic heart disease, appendicitis or renal colic to name a few examples. However, as health care professionals, we rarely use these very same principles for the assessment of pain in cancer patients. The use of these principles can tell us so much about the pain and provide a guide to the best possible treatment options without the need to perform invasive tests (table 3). Pain assessment needs to occur repeatedly and at regular intervals throughout the treatment of a patient and most importantly with any new report of pain.

Components of Pain Assessment

From the work of Portenoy et al. [4], it was noted that many lung cancer patients have more than one pain. It is essential to ask a patient about each component of their pain. It is possible from these individual components to identify pain syndromes, based, for instance, on the distribution of the pain together with the character and radiation. However in seeking such patterns, a clinician must be careful to ensure that a total assessment is performed (table 4).

A. *Intensity:* Physicians and nurses tend to underestimate pain intensity, especially when it is severe. Patients whose doctors underestimate their pain

Table 4. Important components of the pain assessment	A.	Intensity
	B.	Character
	C.	Location
	D.	Radiation
	E.	Timing
	F.	Correlated factors
	G.	Implications of the pain
	H.	Meaning of the pain

are at high risk for poor pain management and compromised function. Communication about pain is greatly aided by having the patient use a scale to report pain severity. A simple rating scale ranges from 0 to 10, with 0 being 'no pain' and 10 being 'pain as bad as you can imagine'. Used properly, pain severity scales can be invaluable in titrating analgesics and in monitoring for increases in pain with progressive disease. Daut et al. [6] used a numeric rating index (0–10) to develop the Brief Pain Inventory (BPI). With this tool, patients self-rate their pain (pain worst, pain average, pain least and pain now) along with a self-assessment of the interference of pain with everyday functions such as activity, mood and relationships with others. The BPI has been validated for cancer patients in several languages and has been used extensively in pain research. Based on continued work with this tool in cancer patients, Serlin et al. [7] have been able to define three levels of pain. Mild pain ('pain worst' score of 0–4) is often well tolerated with minimal impact on a patient's activities. At 'pain worst' scores of 5–6, patients experience some disruption in these daily activities. However, there is a threshold beyond which pain is especially disruptive and is generally reached when the 'pain worst' score is 7 or more on a 0–10 scale. At this level, pain becomes the primary focus of attention and prohibits most activity not directly related to pain. While it may not be possible to totally eliminate pain, reducing its severity to 4 or less should be a minimum standard of pain therapy.

B. *Character:* Verbal descriptors of pain used by the patient may help in establishing the etiology of pain. It is important that a physician be aware of the diversity of terms used by patients. Often a patient will deny that they have pain but that the 'pulling' sensation that they are experiencing in their back would rate as a 7 out of 10 on a 0–10 pain scale. A common descriptive term used by patients is that of an 'ache'. Other terms used include pressure, tightness, burning, tingling, numbness and electric-shock-like pain. A patient's description of a shooting pain down his/her arm would suggest that there is a neuropathic component of the pain.

C. *Location:* The discrimination between generalized versus localized pain is an important one in the consideration of both diagnosis and treatment options. It is equally important to remember that pain can be distributed over dermatomal patterns. This is as important in cancer-related pain as it is when used in the diagnosis of appendicitis (periumbilical pain) or gallbladder disease (right shoulder tip pain). Patients may present with knee pain with no evidence of pathology in that joint. The pain may in fact be referred pain (not radiating) from L3 area in the spine or from disease associated with the hip. Localized pain may be best managed with systemic analgesic agents together with a localized therapy, e.g., radiotherapy or nerve block. If adequate analgesia is obtained through such an intervention, the analgesic agent may consequently be diminished or stopped. Disseminated pain is usually best managed with analgesic agents but a particularly painful location in a patient with disseminated pain may need specific treatment, e.g. radiotherapy.

D. *Radiation:* The radiation of a patient's pain can be a crucial factor in the diagnosis of a pain syndrome. The description of pain in the lower back radiating down both legs tells much about the diagnosis especially if coupled with the verbal description of 'shooting' or like an 'electric shock' suggesting a neuropathic component.

E. *Timing:* Pain may be breakthrough in nature, a term used to describe the transitory flare of pain in the setting of chronic pain managed by opioids. Portenoy and Hagen [8] defined this more specifically as a 'transitory increase in pain to greater than moderate intensity (i.e., severe pain) which occurred on a baseline pain of moderate intensity or less (i.e. moderate or mild pain).' Such pain is also called 'incident pain' when it occurs in association with a specific activity.

F. *Correlated factors:* Specifically defining these correlated factors can help greatly in establishing a diagnosis. Does coughing make low back pain worse and send it shooting down the back? Does straining make a headache worse? It is important to elicit the medications (not always analgesics) that a patient has been taking for the pain, together with the impact of these medications. This information helps in the assessment of pain severity and together with the use of equianalgesic dose tables will allow the adequate dose of other analgesic agents to be prescribed.

G. *Implications of the pain:* Pain may limit a patient's activity that may already be dramatically effected by other cancer-related symptoms. Therefore, it is imperative to document the effect of the pain on all patients' lives. This is more so in the elderly who, because of decreased reserves, may be more severely compromised by pain. When pain is of moderate or greater severity, we can assume that it has a negative impact on the patient's quality of life. That impact, including problems with sleep and depression, must be evaluated. The number of hours the patient is now sleeping compared with the last pain-free interval, difficulties with sleep onset, frequent interruptions of sleep, and/or early morning awakening suggest the need for appropriate pharmacological intervention. This may include the addition of a low-dose antidepressant at bedtime. Equally important is an assessment of how pain interferes with a patient's ability to interact with others. The BPI, while used initially as a research tool, provides assessment of pain interference and has been found to be useful in clinical practice [9].

Significant depression should be treated through psychiatric or psychological consultation, especially if it persists in the face of adequate pain relief. Just as patients hesitate to report severe pain, they may hesitate to report depression. Having the patient report depression or tension on a scale of 0–10 may help overcome some of this reluctance. It is also known for physicians to misdiagnose pain as depression or anxiety. Patients who are cognitively impaired, particularly those with agitation, may be extremely difficult to assess. In these patients, the differentiation between agitated delirium and pain may be extremely difficult. Patients in whom pain was well controlled before the development of delirium are unlikely to be agitated due to uncontrolled pain. Frequent discussions between various health care professionals and the patient's family are often required.

H. *Meaning of the pain:* The implication of pain to a cancer patient is a very important part of the pain assessment. A patient's understanding of an initial minor increase in pain may dramatically influence his/her perceived severity of the pain that could then result in the need to escalate the dose of an opioid. Does the pain mean the recurrence of a malignancy that the patient thought had been 'cured'? Does the worsening of a pain indicate that the patient's disease is progressing despite chemotherapy or is it related to the treatment? Will the pain mean more investigations, many of which are uncomfortable and painful? Considerations such as these may lead to the underreporting of pain in some patients while in others will increase the anxiety associated with pain.

Assessment of Other Symptoms

While pain has been extensively studied, there has been less work conducted on the assessment of other symptoms. The Edmonton Symptom Assessment System (ESAS), developed in a palliative care setting, consists of nine visual analogue scales that include pain, shortness of breath, nausea, depression, activity, anxiety, well-being, drowsiness and appetite [10]. Each visual analogue scale consists of a 100-mm line representing the lowest symptom intensity (e.g. 'no pain') to the worst possible degree of symptom intensity (e.g. 'worst possible pain'). Whenever the assessment is performed, the score is then transferred to a bar graph that allows the staff to visualize patterns of symptom control and symptom expression over time. It allows for the interpretation of visual patterns associated with the predominance of symptoms, and can also be used for quality control by documenting the characteristics of patients admitted to different areas of a comprehensive palliative care program. The European Organization for Research and Treatment of Cancer has introduced a quality-of-life instrument [11], the QLQ-C30, which incorporates the assessment of some symptoms as well as assessment of function on five scales. The tool has been validated in 354 patients with unresectable lung cancer throughout the world.

Fatigue is a major symptom in cancer patients and efforts are currently underway to address the assessment of fatigue and the cause of this debilitating symptom in cancer patients. The Multidimensional Fatigue Inventory has been developed as a tool for the measurement of fatigue in cancer patients [12]. This index studies four dimensions of fatigue and asks patients to rate their level in a series of boxes. A Brief Fatigue Inventory is currently under development at the University of Wisconsin. This is based on the same principles as the BPI [6] and has patients rate their assessment of fatigue together with the degree to which it has interfered with other components of their lives.

Further Assessment: Examination and Imaging

The physical examination should include a full disease assessment as well as a careful examination of the painful area or areas. This examination should be as least distressing for the patient as possible while respecting that analgesics may mask some of the signs necessary to fully make a diagnosis. Equally the treatment of pain should not have to wait for the full diagnosis of a patient's disease. A patient is often transferred to a radiology department with inadequate pain management, only to return with an inadequate procedure because of poor analgesia. Pain management should commence as soon as possible in the treatment of a cancer patient and should be based soundly on the suspected pathophysiology of the disease together with the pharmacology of the analgesic agents. The use of short-acting opioids administered intravenously can provide this balance and ensure both patient and physician are comfortable.

The decision as to how extensively to investigate pain depends on a careful balance between what we know of the pain etiology in relation to the patient's disease stage and his/her associated prognosis. It may be that sudden severe back pain represents collapsed vertebrae together with an impending spinal cord compression and that investigation with MRI needs to be performed to identify the lesion. A more gradual increase in pain in the same area may represent progressive bony disease that can be managed with an increase in analgesic agents without further investigation, although cord compression must always be a consideration. The absence of neurological signs does not exclude an impending spinal cord compression. A thorough understanding of the pathophysiology of the disease process is imperative.

Bone metastases are a common cause of pain in disseminated lung cancer and lesions may be detected on plain x-ray. Limb pains in association with weight bearing may indicate the posibility of an impending fracture, the occurrence of which may severely debilitate a patient and which may be prevented with orthopedic and radiotherapy intervention. However, since painful metastases can occur without changes on plain x-ray, bone scan may be a more effective investigative procedure. CT or MRI scanning is useful in the evaluation of retroperitoneal, paravertebral, and pelvic areas as well as the base of the skull. Myelography may also be necessary in determining the cause of pain. Diagnostic nerve blocks can provide information concerning the pain pathway and can also determine the potential effectiveness of neuroablative procedures. It is important to remember that medical intervention is not always welcomed by patients. One of the fears that patients report is of pain associated with investigations and procedures. Patients may not report pain or they may underplay the severity of their pain for fear of being ordered yet another procedure of x-ray [13].

Symptom Management in Lung Cancer

Direct treatment of symptoms can be very beneficial to patients and should be directed not only at those near to death but also to those receiving antitumor therapy. Therapy directed at the tumor may not always bring about a cure but may cause a decrease in symptoms, e.g. steroids and radiotherapy for brain metastases. With all treatment offered, both supportive and antitumor, a balance must be achieved between benefit and toxicity.

The Management of Pain
The principles of cancer pain management have been reviewed in a number of forums [14, 15] that are widely available in paper and electronic forms[1] in English and other languages. Most commonly these principles are divided into pharmacological and nonpharmacological treatments.

The World Health Organization (WHO) has recommended the use of the analgesic ladder for the treatment of cancer pain [14]. With this comes the

---

[1] http://www.stat.washington.edu/TALARIA/TALARIA.htal.

---

basic tenants for the management of cancer pain; by the ladder by mouth, by the clock and for the individual. There has been some confusion concerning the best way to use the ladder. The WHO recommends that cancer pain treatment progress through the various steps of the ladder to achieve relief from cancer pain [14]. This has been tested and found to be a satisfactory means of providing pain relief [16]. However, patients' pain intensity ratings at the various steps were not reported nor were the duration of inadequate analgesia prior moving up the next step of the ladder. Zech et al. [17] reviewed their experience with pain relief in 2,266 cancer patients, 74% of whom were on step II or III opioids at the time of admission to their pain service. While useful analgesia was obtained using the ladder (efficacy of pain relief was 'good' in 76%), it is difficult to conclude from this study that all patients should be commenced at step 1 of the ladder. Twenty-five percent of patients of step 1 analgesics had pain intensities that were rated moderate to severe using a verbal rating scale, supporting concerns that by progressing through each step, many patients were without pain relief for some time.

The recent AHCPR guideline [15] for the management of cancer pain in the USA recommend use of the WHO ladder for cancer pain management but that patients be treated according to the severity of their pain. Mild pain can be treated at step 1 with analgesics such as nonsteroidal anti-inflammatory drugs (NSAIDs) or acetaminophen. For moderate pain, opioids are commenced with either codeine, or low doses of oxycodone or morphine. Severe pain is treated at step 3 with full doses of opioids including morphine, oxycodone, hydromorphone or fentanyl. Consideration of the use of adjuvant medications is recommended in all cases and should include the use of NSAIDs together with opioids in the case of bony metastases. Adjuvant medications may result in a decrease in opioid dose with a concomitant decrease in side effects. This is the subject of further research.

Physicians have a vast array of analgesic agents available for use with their patients. It is essential that physicians treating cancer pain understand the pharmacology of two to three drugs from each of the steps of the cancer pain ladder. The dose of pain medications is titrated upwardly until either a patient's desired pain relief or until *unmanageable* side effects are reached. In the case of NSAIDs and acetaminophen, dose escalation will be limited by either side effects or the expectation of side effects. Opioids in their own right do not have a fixed ceiling in their dose, the highest dose being that which provides analgesia or results in *unmanageable* side effects. When opioids are combined with acetaminophen or aspirin such as in codeine/acetaminophen or oxycodone/acetaminophen formulations, the dose-limiting component of these preparations is the total daily dose of acetaminophen or aspirin.

Analgesic agents are recommended to be taken on a regular around-the-clock basis, not on a 'prn' or 'as needed' basis. This relates to the need to maintain adequate concentrations of an opioid in the body and is more likely to maintain a patient in a pain-free state [18]. Rescue or 'as needed' medications should be ordered for all patients and should be used for episodes of breakthrough pain and during periods of titration (either upward or downward titration) of analgesic agents. Current recommendations for rescue doses range from 5 to 15% of the daily dose [8] (or 10–30% of the 12-hourly dose). Immediate release oxycodone and morphine are most often used for the treatment of breakthrough pain, although fentanyl delivered in an oral transmucosal formulation has recently been found to be effective in the treatment of breakthrough pain [19].

The initial treatment and titration with opioids should take place with immediate-release preparations of opioids [20]. The prolonged absorption of a sustained-release product may result in prolonged side effects in a patient who is receiving opioids for the first time. However, a patient who is taking Percoset® is not opioid naive and may be changed to a sustained-release product. Twelve Percoset® tablets are the equivalent of 60 mg/day of oral morphine. Oxycodone, the active ingredient in Percoset®, is a drug that has an analgesic activity similar to morphine and which is now available as a sustained-release product. The use of sustained-release products has made twice daily dosing of analgesics a reality for cancer patients and once-a-day morphine products are also now available (Kadean®/Kapanol®). There is however no statistically significant difference in the side effects and analgesia between immediate-release morphine administered every 4 h and sustained-release morphine administered twice daily [21]. Patients taking sustained-released morphine who achieve analgesia that is not sustained for the full dosing interval may benefit from dosing 3 times per day. Methadone, another oral opioid more commonly associated with drug withdrawal programs, is a cheap and effective alternative for the treatment of cancer pain. Methadone has a long half-life and therefore dose escalation must proceed cautiously, especially in the elderly in order to reduce the occurrence and severity of side effects. Previous studies have suggested that the conversion ratio of morphine to methadone is 3:2. However, recent data would suggest a 10:1 conversion ratio with chronic dosing of methadone [22].

Ideally, pain medications should be given by mouth. However, some patients cannot tolerate the oral route. In the study of Zech et al. [17], approximately 80% of the 2,118 patients were managed with oral medications throughout their illness. This decreased to 50% of 864 patients being cared for in hospital at the time of death. Alternative methods of drug delivery therefore need to be considered in those in whom oral administration is not possible. Many people use the sublingual route feeling that this provides rapid analgesia. There is increasing evidence that the sublingual administration of morphine provides no benefit over oral administration. Peak plasma concentrations of morphine occur later and at lower levels following sublingual administration than for oral administration [23]. It may also be preferable to use morphine solutions rather than sublingual tablets that have a bitter taste. The rectal administration of opioids is not limited to drugs for which there is a suppository formulation; sustained-release tablets administered rectally provide effective analgesia and have a similar pharmacokinetic profile to the same formulations given orally.

The transdermal delivery of opioids is currently limited to fentanyl. The rate of delivery of fentanyl (μg/h) is dependent on the surface area of the patch applied, with rates ranging from 25 to 100 μg/h. When applied to the skin, fentanyl passively diffuses across a concentration gradient into the subcutaneous fat and then more slowly into the plasma resulting in sustained plasma concentrations. At approximately 72 h, the concentration of drug in the fat equilibrates with the concentration of drug in the patch and a new patch needs to be applied to a new area of the patient's trunk. Current recommendations are that patients should be stabilized on oral opioids prior to starting transdermal fentanyl and that dose changes should not be made more frequently than every 72 h [24]. As always, patients need to be ordered a short-acting opioid, such as immediate-release morphine or oxycodone, for breakthrough pain; oral transmucosal fentanyl citrate, currently approved for preoperative sedation in children, may be a useful rescue medication for cancer patients.

The parenteral administration of opioids may be necessary in those who cannot swallow, who have intractable side effects or in whom rectal delivery is not desirable. Subcutaneous infusions have been extensively used in Canada [25] and Australia but are not commonly used in the USA, possibly because many cancer patients have intravenous ports. The steady plasma concentrations of opioids, resulting from either intravenous or subcutaneous infusions, may result in a diminution of side effects and therefore an optimization of analgesia. Only a small percentage of cancer patients (2–5%) will require interventions or the direct delivery of opioids to the CNS [16]. The cost differential of the different treatment options needs to be considered. Patients with unmanageable side effects may benefit from epidural or intrathecal administration of opioids. Approximately one tenth of the intravenous dose of an opioid needs to be administered epidurally and one hundredth the dose administered intrathecally. These procedures are costly, needing catheters and pumps to deliver drug and are of questionable cost effectiveness in patients with a short life expectancy.

The intramuscular injection of opioids is to be avoided. Apart from being painful for patients, absorption following injection is erratic and in most cases results in an analgesic effect that parallels the oral administration of an equivalent dose of the same drug. One opioid commonly given by intramuscular injection is meperidine or pethidine (Demerol®). Meperidine is in fact a drug that has very few indications in the treatment of pain. Meperidine has short-acting analgesic activity (one tenth that of morphine) and is mostly prescribed at subtherapeutic doses (25–75 mg every 3–4 h i.m.). It is metabolized to normeperidine, a toxic metabolite that accumulates with repeated administration and can result in many side effects including hallucinations and seizures [26]. If used at all, meperidine should be used for no more than 48 h and with a dose limitation of 600 mg/day [15]. This dose limitation should be 450 mg/day in the presence of renal impairment that occurs more commonly in the elderly. Given these limitations, meperidine is not recommended for the routine treatment of pain.

A small minority of patients who have alcoholism or drug addiction may request analgesics for psychological effects. This is unlikely to occur in patients without a clear history of severe addictive behavior. Patients who are recovered alcoholics or drug abusers may be difficult to treat because of their resistance to taking analgesics. Although their care is more complex, lung cancer patients with drug or alcohol addiction or a history of addiction, should never be denied appropriate pain medications. If drug addiction is suspected, the patient should be presented with these suspicions and agreement should be made about the use of opioids for the management of pain as opposed to the alteration of mood. The use of long-acting opioids or continuous infusions is preferable to short-acting opioids or patient-controlled analgesia. The writing of prescriptions by a single physician can simplify the negotiation process with such patients.

While opioids are the mainstay of cancer pain management, the use of adjunct therapy is recommended by both the WHO [14] and AHCPR guidelines [15]. NSAIDs are particularly useful in the management of metastatic bone pain. While many prefer to use agents such as ibuprofen and naproxen, aspirin is equally effective in the management of bone pain. Steroids may be also useful in the management of bone pain but may cause unwanted side effects. Steroids are particularly useful in the management of painful liver metastases, where they act by reducing pressure on the liver capsule, the cause of hepatic pain. Antidepressants such as amitriptyline and desimpramine are useful adjuncts for all cancer pain but particularly important for the treatment of neuropathic or nerve pain. Antiepileptics such as carbamazepine have also been used as second-line adjuncts for neuropathic pain and the effectiveness of gabapentin, a newer antiepileptic, is currently being investigated.

The Management of Dyspnea

Dyspnea refers to the sensation of 'bad breathing' (dys pneo), a symptom that up to 75% of lung cancer patients experience. Some 70% of patients close to death in hospices experience dyspnea and in 24% of these, no underlying cardiac cause was documented as a cause of their dyspnea [27]. 41% of palliative care patients have been documented to have dyspnea and 46% describe it as being of moderate to severe intensity [28].

Oxygen Therapy. In clinical practice we associate the symptom of dyspnea with hypoxia but it may be associated with many of the disease manifestations of lung cancer, including superior vena cava syndrome, pleural disease and bronchial obstruction. In the management of dyspnea, one must not forget that there are reversible causes of the symptom that can be managed symptomatically. Many patients may have bronchoconstriction that will respond to β-receptor agonists, methylxanthines and corticosteroids. Steroids are usually useful in the management of obstruction and lymphangitic spread. The relief of obstruction, and consequently of dyspnea, with laser therapy and/or stents may provide very good symptomatic relief to some patients.

Continuous oxygen therapy is most commonly associated with chronic obstructive airway disease (COAD), in which its use reduces pulmonary hypertension. In cancer patients, oxygen is primarily used to reduce dyspnea. Maximal relief has been found in cancer patients who are hypoxic with dyspnea at rest [29]. However, intermittent oxygen can be beneficial for exertional dyspnea. The delivery of oxygen though a concentrator (nitrogen extractor) is preferable to cylinders. The use of nasal cannulae rather than a mask may allow patients to function at a higher level. The flow rate of the oxygen should be directed at patient comfort. Oxygen saturation may assist in titrating levels and arterial blood gas analysis may be necessary in those who are carbon dioxide retainers.

Drug Interventions for Dyspnea. Opioids have been used with great effectiveness in the treatment of acute dyspnea. Morphine is part of the primary treatment in cardiac failure in which it acts as a vasodilator, depresses respiratory rate and alters sensation of dyspnea. Sublingual, oral and intravenous morphine formulations are commonly used in the treatment of cancer-associated dyspnea with good effect. For continuous dyspnea, small doses of long-acting morphine may be effective. Pain is often an associated symptom and the use of opioids will be directed primarily at that symptom. Intermittent (breakthrough) dyspnea can respond to 'prn' opioids. There have however been few studies to elucidate opioid use in these situations. It is always important to ensure whether another therapy (drug or nondrug) may be more successful. The increased escalation of opioid doses may result in excessive sedation beyond that seen when opioids are administered for pain control.

Opioid receptors have been identified in the airways and it has been suggested that nebulized opioids may be useful for the treatment of dyspnea. Advantages of the nebulized delivery include direct action in the airways and a large surface area available to provide rapid and efficient delivery of drug. Many patients with lung cancer are familiar with the use of nebulizers in the treatment of underlying lung disease. Morphine delivered by nebulizer has shown one sixth the difference in peak plasma levels [30] and the delivery of 5 mg by nebulizer is the equivalent of 1.7 mg morphine administered intravenously. Five milligrams of nebulized morphine resulted in a 35% increase in exercise capacity in COAD patients [31]. Despite no double-blind studies in cancer patients to clarify the efficacy of nebulized morphine, its use is increasing within hospice circles. Such studies are needed. Care needs to be taken when administering morphine by nebulizer. Bronchospasm may result and therefore a test dose should be given in the clinic setting. There is some suggestion fentanyl may be a better option as it is associated with less histamine release. Other side effects of nebulized morphine are commensual with equivalent parenteral doses.

Respiratory sedatives are also widely used in the treatment of dyspnea. Few studies have been performed in cancer patients that address the use of compounds such as phenothiazines and benzodiazepines. Promethazine has been shown to reduce dyspnea in COAD patients. Diazepam 5 mg administered to COAD patients resulted in improved sleep and had no effect on nocturnal hypoxia. There are concerns about carbon dioxide retention in COAD patients. Lorazepam is a relatively quick-acting benzodiazepine that can be administered orally, sublingually or intravenously. Midazolam can be administered as an infusion either intravenously or subcutaneously.

The Management of Fatigue

Fatigue in lung cancer patients may be due to many causes including the tumor itself, treatment, anemia, infection or metabolic abnormalities. It is essential, prior to treating fatigue, to give careful consideration to those disease processes that may respond to direct therapy, e.g. infection, hyperkalemia and Addison's disease. It is also essential, prior to the commencing diagnostics and treatment of fatigue that the relevance of the symptom in relation to a patient's disease process, expected lifespan and quality of life be considered.

Correction of Anemia. It has already been suggested that anemia may be overestimated as a cause of fatigue in cancer patients. This relationship may become more evaluable as pharmacological treatment of anemia using synthetic erythropoietin becomes more widespread. Among 118 cancer patients not receiving chemotherapy, erythropoietin resulted in a 2.4% increase in hematocrit, which was significantly greater (-0.1%) than the change seen with placebo [32]. There was, however, no statistically significant difference in the absolute hematocrit levels between the two groups. Although pretreatment erythropoietin levels were a significant predictor of response in this group, there was no significant difference in the transfusion requirements of patients receiving placebo or erythropoietin over the 8 weeks of the study. One has to question the role of synthetic erythropoietin administration in this group. For those receiving chemotherapy (with and without cisplatin), there was a significant change in hematocrit with the administration of erythropoietin in the first month, but no significant change in transfusion requirements. Transfusion requirements decreased significantly in the erythropoietin group during the second and third months of treatment but only in those patients receiving cisplatin.

One difficulty with the results of these studies is the way that the results are presented. QOL data are only presented in the responders, that is in those who had a greater than 6% increase in their hematocrit. While these increases were significantly different from the pooled placebo data, the change in visual analog scores for those treated with erythropoietin compared to the placebo group was only 10 mm (out of 100 mm) for energy level, daily activities and overall QOL. It is difficult to appreciate how meaningful this 10-mm change would be to patients. A 10% decrease in fatigue scores may have little impact on a patient's QOL if similarities are drawn with the measurement of pain [7]. One would also suspect that the inclusion of nonresponders in these results may result in a nonsignificant finding.

Nutrition. There is evidence that aggressive improvement in nutrition increases survival, tumor response, or treatment toxicity among those with advanced cancer. There is no evidence that aggressive nutritional therapy improves the quality of life of these patients [33]. Parenteral feeding of cancer patients has little impact on fatigue [34].

Drug Interventions. The cessation of other drug therapies that cause asthenia is recommended. Difficulties arise when these therapies are used with curative or life-prolonging intent. Steroids may need to be tapered slowly in patients who experience fatigue following longer use of these drugs.

Supplemental steroid therapy can ameliorate fatigue and may be particularly indicated during some anticancer treatments, such as high-dose ketoconazole therapy [35]. Moertel et al. [36] found a significant improvement in appetite and strength in patients after 2 weeks of dexamethasone treatment (0.75 and 1.5 mg qid), compared with placebo. However, this improvement disappeared after 4 weeks of treatment. Bruera et al. [37] observed that methylprednisolone caused a rapid improvement in activity level in a double-blind setting but this improvement was not sustained over a 3-week period. In another randomized, placebo-controlled double-blind trial, Bruera et al. [38] found that 1 week of megesterol acetate resulted in an improvement in appetite, caloric intake, nutritional status and level of energy. This improvement may have been a nonspecific steroid effect of megesterol, rather than a specific action on fatigue. While steroids have been used to treat the symptoms of asthenia, a dilemma arises if steroid-induced myopathy occurs. The treatment options in this setting are reduction of steroid dose, changing to a nonfluorinated steroid, alternate-day dosing, and isometric exercise [39].

Amphetamines may be useful in patients who have treatment-associated fatigue. Mazindol had little effect on the activity score of cancer patients [40], but methylphenidate resulted in a significant improvement in the level of activity [41]. In the latter study, patients were taking large doses of opioids and the methylphenidate may have had an indirect effect by improving pain control or opioid-induced sedation.

Exercise. Exercise has been helpful in improving symptoms in women receiving chemotherapy [42]. A structured 10-week exercise program reduced fatigue, stabilized weight and reduced nausea in stage II breast cancer patients treated with chemotherapy. However, caution must be urged to ensure that the exercise level is appropriate for the patient and does not result in the development of fatigue [43]. Exercise is often very difficult in lung cancer patients because of underlying respiratory disease. The use of an activity diary that also records patient self-assessment of fatigue may be useful [44]. This allows patients to make the most of high-energy times and allows clinicians to schedule dose reduction and/or a break from therapy, if possible within the goals of treatment.

The Role of Antitumor Therapy in Palliation

Anticancer treatment is often divided into active and palliative treatments. This demarcation is often blurred, and is particularly so in the treatment of non-small cell lung cancer (NSCLC). It is probably more appropriate that treatment be classified according to the intention of treatment: curative, prolongation of survival or palliative. Many of the antitumor therapies used in lung cancer have been found to extensively palliate patients' symptoms.

Radiotherapy (XRT)
XRT plays a significant role in the palliative treatment of lung cancer patients. There is considerable evidence that single-dose or short-term XRT provides as effective pain relief as does longer courses of therapy for many forms of primary tumors as well as metastatic disease [45]. Currently, both in Canada and Great Britain, single-dose and short-course XRT have become

the standard for palliative XRT. For bone metastases, single-fraction XRT is used for palliation as opposed to multiple-fraction treatment. For lung cancer, 10-fraction (20 Gy) XRT treatments have become the norm for palliation.

The distinction between palliative and curative XRT warrants attention. Palliative XRT, via mechanisms not fully understood, serves only to alleviate symptoms. In curative, or 'radical' XRT, the intent is to eradicate tumor cells with minimal healthy tissue damage. With curative intent, high doses (50–60 Gy) are spread out over 4–6 weeks and fractions are kept relatively small to minimize toxicity and complications that may arise months or years later from exposure to large single doses of XRT. In the case of hospice patients, the latter consideration should be of little concern because all patients admitted to hospice in the USA are deemed to have a life expectancy of 6 months or less. In this patient group, short courses of higher dose per fraction XRT are especially appropriate.

This is not the case in the USA, however. It may be that given the parameters described above, many hospice patients are receiving treatment that is more curative in nature than palliative. In a national study for 1984–1985, 53.8% of the patients in the USA received XRT of palliative intent that lasted 15 day or longer [46]. There remains debate in the USA, represented by the conflicting conclusions of the Radiation Oncology Group's (RTOG) study, 'The Palliation of Symptomatic Osseous Metastases' and a subsequent re-analysis of the data as to the best treatment protocol. The original analysis suggests that single-fraction radiotherapy is as effective as multiple-fraction regimens for palliation of bone metastases [47]. The re-analysis raised questions concerning the use of pain scores in the original report and concluded in favor of multiple fraction treatments [48]. It is suggested in the paper that financial considerations may have stalled the acceptance of short-course therapy for bone metastases.

The role of XRT in inoperable NSCLC, too advanced for 'curative' treatment, has also been studied in a multicenter study by the Lung Cancer Working Party of the Medical Research Council [2]. The majority of patients required palliative treatment for major symptoms related to intrathoracic tumor. For many of these patients in whom the disease was considered too advanced to be treated with curative intent, the patient was administered 30 Gy in 10 fractions or the biologically equivalent dose of 27 Gy in 6 fractions. In a randomized study the role of 2 fractions of 7.5 Gy, given 1 week apart, was compared with the standard therapy. Palliation and toxicities were assessed in 369 patients. Equal numbers of patients from both groups received treatment with analgesics, steroids and bronchodilators. Palliation was equal in both groups and the median time of palliation was 50% or more of survival. Palliation ranged from 69 to 146 days in the two-fraction and from 71 to 140 days in the multiple-fraction arm. Tumor response was similar in both groups and there was no survival difference between the two groups. As with many other lung cancer studies, the level of physical activity on admission was the factor with the greatest effect on the duration of survival. The authors concluded that for patients with inoperable disease and poor performance status, short-course palliative radiotherapy was a useful option.

There are several clear advantages to short-course palliative XRT. One such benefit may be measured in terms of 'opportunity cost' as conceived by Munro and Sebag-Montefiore [49]. For the patient with a limited lifespan, each day spent traveling to and from the hospital for therapy represents a relatively large portion of their remaining lifetime. The cost is measured in lost activities and interests the patient might have otherwise pursued. Additionally, the financial benefit of short-course XRT compared to longer treatments is of interest, especially to the insurer. Under the current Hospice Medicare agreement in the USA, the hospice provider is required to pay for all outpatient therapy, which in many cases includes radiotherapy. The use of short-fraction radiotherapy may result in financial savings that could be reallocated to other patient services with no diminution of pain control for these patients.

Chemotherapy

There is no question that small cell lung cancer is generally responsive to chemotherapy and that symptoms can be dramatically improved with the use of antitumor drugs. Patients with a poor performance status may have a dramatic improvement in activity and quality of life with the use of chemotherapy. There has been ongoing research to minimize the toxicities of schedules associated with such drugs. It has been suggested that etoposide, an active agent against small cell lung cancer, can be administered orally as a single agent and provides effective palliation against the disease. It has been considered as therapy in the elderly and for those who have experienced a relapse of their disease.

The role of chemotherapy in NSCLC is more controversial and has been considered elsewhere. Because chemotherapy has achieved poor response rates for NSCLC, randomized studies comparing chemotherapy with 'best supportive care' have been undertaken. Chemotherapy has shown an improvement in survival but at a cost with the authors of these studies reporting severe to life-threatening toxicity in many of patients. From a financial perspective, chemotherapy has in some cases found to be as cost effective as best supportive care [50]. A number of meta-analyses have now examined the role of best supportive care with or without the addition of chemotherapy [51–54]. Early studies using long-term alkylating agents and vinca alkaloids and etoposide favored best supportive care in terms of survival. However, when cisplatin-based regimens are taken alone (778 patients), chemotherapy shows a survival

Table 5. Physician selection of treatment modalities for stage IIIB and stage IV NSCLC [56]

Stage	No immediate treatment	Radiation	Chemotherapy
IIIB	17%	65%	17% (+radiation)
IV	79%	–	19%

advantage over best supportive care. The Non-Small Cell Lung Cancer Collaborative Group [55] meta-analysis showed a benefit of chemotherapy, with a hazard ratio of 0.73 ($p<0.001$) into a reduction in the risk of death of 27%, equivalent to an absolute improvement in survival of 10% at 1 year or an increased median survival of 1.5 months. Many of the studies analyzed involved continuous chemotherapy until disease progression or toxicity and did not address the issue of short-term chemotherapy. None of the studies addressed the issue of quality of life that must be an important consideration in the treatment of any patient with advanced cancer. Equally the studies excluded patient's with a poor performance status. Both quality of life and the use of fixed short-term courses of chemotherapy need to be addressed in prospective randomized studies. The later is currently being addressed in a four-arm study of the Eastern Cooperative Oncology Group that includes new agents such as taxotere and gemcitabine. Unfortunately there is no 'best supportive care arm' in this study nor is quality of life measured.

Many physicians who treat NSCLC believe that chemotherapy does not have a role to play in NSCLC [55]. When asked what treatment they would believe as the value of treatment of various stages of NSCLC, most did not include chemotherapy as an option (table 5). This study had been conducted in both 1986 and 1994, showing that recent studies comparing best supportive care to chemotherapy had little impact on treating physicians' personal attitude to chemotherapy for NSCLC. The impact of an increased survival of 1 month must be questioned. It is imperative, however, that patients be considered for clinical studies addressing the role of chemotherapy.

Ethical Dilemmas Near the End of Life

Ethical considerations are important in the palliation of symptoms related to lung cancer. One fear expressed in the use of opioids is that of respiratory depression as many lung cancer patients have underlying COAD. Respiratory depression is an uncommon side effect with the careful titration of opioids and should therefore not limit the treatment of pain. One can see that opioids may benefit the symptom of dyspnea which if anything should encourage the use of opioids in lung cancer patients. Clinically it appears that most patients develop tolerance to many of the side effects of opioids including that of respiratory depression. Physicians may assess sedation or drowsiness as being equivalent to respiratory depression. Many patients who have endured severe pain may be sleep deprived and with adequate analgesia, require some days to 'catch up'.

In a very small number of patients, symptoms may be difficult to control with the methods outlined above. In these cases it may be necessary to sedate patients in order to relieve their suffering. This is endorsed by sound medical and ethical principles. The rule of double effect is also an important consideration. Following this principle, and appropriate dose of opioids may be administered to relieve pain even if as a secondary effect death is hastened. Critical to this is the issue of appropriate dose. The titration of opioids rather than the use of large boluses is critically important in order to ensure that the intention is indeed symptom relief and not the hastening of death.

Conclusion

Lung cancer is still, and will continue to be, a major cause of death in our community. With appropriate assessment and treatment, many of the symptoms of lung cancer patients can be successfully palliated. The importance of palliating these symptoms in all patients, regardless of the stage of disease, must be recognized by clinicians involved in the treatment of lung cancer.

References

1   American Cancer Society: Cancer Facts and Figures. Atlanta, American Cancer Society, 1996.
2   Bleehen NM, Girling DJ, Fayers PM, Aber VR, Stephens RJ: Inoperable non-small cell lung cancer: A Medical Research Council randomised trial of palliative radiotherapy with two fractions or ten fractions. Br J Cancer 1992;63:265–270.
3   Muers MF, Round CE: Palliation of symptoms in non-small cell lung cancer: A study by the Yorkshire Regional Cancer Organization Thoracic Group. Thorax. 1993;48:339–343.
4   Portenoy RK, Miransky J, Thaler HT, Hornung J, Bianchi C, Cibas-Kong I, Feldhamer E, Lewis F, Matamoras I, Sugar MZ: Pain in ambulatory patients with lung or colon cancer. Prevalence, characteristics, and effect. Cancer 1992;70:1616–1624.
5   Von Roenn JH, Cleeland CS, Gonin R, Hatfield AK, Pandya KJ: Physician attitudes and practice in cancer pain management. A survey from the Eastern Cooperative Oncology Group. Ann Intern Med 1993;119:121–126.
6   Daut RL, Cleeland CS, Flanery RC: Development of the Wisconsin Brief Pain Questionnaire to assess pain in cancer and other diseases. Pain 1983;17:197–210.
7   Serlin RC, Mendoza TR, Nakamura Y, Edwards KR, Cleeland CS: When is pain mild, moderate and severe? Grading pain severity by its interference with function. Pain 1995;61:277–284.

8 Portenoy RK, Hagen NA: Breakthrough pain: Definition, prevalence and characteristics. Pain 1990; 41:273–281.

9 Cleeland CS, Ryan KM: Pain assessment: Global use of the Brief Pain Inventory. Ann Acad Med Singapore 1994;23:129–138.

10 Bruera E, Kuehn N, Miller MJ, Selmser P, Macmillan K: The Edmonton Symptom Assessment System: A simple method for the assessment of palliative care patients. J Palliat Care 1991;7:6–9.

11 Aaronson NK, Ahmedzai S, Bergman B, Bullinger M, Cull A, Duez NJ, Filiberti A, Flechtner H, Fleishman SB, de Hacs CJM, Kaasa S, Klee M, Osaba D, Razavi D, Rofe PB, Schraub S, Sneeuw K, Sullivan M, Takeda F: The European Organization for Research and Treatment of Cancer QLQ-C30: A quality-of-life instrument for use in international clinical trials in oncology. J Natl Cancer Inst 1993;85:365–376.

12 Smets EM, Garssen B, Bonke B, De Haes JC: The Multidimensional Fatigue Inventory psychometric qualities of an instrument to assess fatigue. J Psychosom Res 1995;39:315–325.

13 Ward SE, Goldberg N, Miller-McCauley V, Mueller C, Nolan A, Pawlik-Plank D, Robbins A, Stormoen D, Weissman DE: Patient-related barriers to management of cancer pain. Pain 1993;52: 319–324.

14 World Health Organization: Cancer pain relief and palliative care: Report of a WHO Expert Committe. Geneva, WHO 1990.

15 Jacox A, Carr BD, Payne R: Management of Cancer Pain. Clinical Practice Guideline No 9. AHCPR Publ No 94-0592. Rockville, Agency for Health Care Policy and Research/US Department of Health and Human Services, Public Health Service, 1994.

16 Ventafridda V, Oliveri E, Caraceni A, Spoldi E, De Conno F, Saita L, Ripamonti G: A retrospective study on the use of oral morphine in cancer pain. J Pain Symptom Manage 1987;2:77–81.

17 Zech DFJ, Grond S, Lynch J, Hertel D, Lehmann KA: Validation of World Health Organization guidelines for cancer pain relief – A 10-year prospective study. Pain 1995;63:65–76.

18 Lipman AG: Opioid analgesics in the management of cancer pain. Am J Hospice Care 1989;6:13–23.

19 Clearly JF: A randomised double-blind, placebo-controlled study for management of breakthrough pain: Oral transmucosal fentanyl citrate versus placebo. Proc Am Soc Clin Oncol, 1997, abstr 179.

20 Portenoy RK: Pan management in the older cancer patient. Oncology 1992;6:S86–S98.

21 Walsh TD, MacDonald N, Bruera E: A controlled study of sustained release morphine sulfate tablets in chronic pain from advanced cancer. Am J Clin Oncol 1992;115:268–272.

22 Crews JC, Sweeney NJ, Denson DD: Clinical efficacy of methadone in patients refractory to other mu-opioid receptor agonist analgesics for management of terminal cancer pain. Case presentations and discussion of incomplete cross-tolerance among opioid agonist analgesics. Cancer 1993;72: 2266–2272.

23 Osborne R, Joel S, Trew D, Slevin M: Morphine and metabolite behavior after different routes of morphine administration: Demonstration of the importance of the active metabolite morphine-6-glucuronide. Clin Pharmacol Ther 1990;47:12–19.

24 Payne R, Chandler S, Einhaus M: Guidelines for the clinical use of transdermal fentanyl. Anticancer Drugs 1995;6(suppl 3):50–53.

25 Macmillan K, Bruera E, Kuehn N, Selmser P, Macmillan A: A Prospective comparison study between a butterfly needle and a Teflon cannula for subcutaneous narcotic administration. J Pain Symptom Manage 1994;9:82–84.

26 Kaiko RF, Foley KM, Grabinski PY, Heidrich G, Rogers AG, Inturrisi CE, Reidenberg MM: Central nervous system excitatory effects of meperidine in cancer patients. Ann Neurol 1983;13:180–185.

27 Reuben DB, Mor V: Dyspnea in terminally ill cancer patients. Chest 1986;89:234–236.

28 Curtis EB, Krech R, Walsh TD: Common symptoms in patients with advanced cancer. J Palliat Care 1991;7:25–29.

29 Bruera E, de Stoutz N, Velasco-Leiva A, Schoeller T, Hanson J: Effects of oxygen on dyspnoea in hypoxaemic terminal-cancer patients. Lancet 1993;342:13–14.

30 Chrubasik J, Wust H, Friedrich G, Geller F: Absorption and bioavailability of nebulized morphine. Br J Anesth 1988;61:228–230.

31 Young IH, Daviskas E, Keena VA: Effect of low-dose exercise endurance in patients with chronic lung disease. Thorax 1989;44:387–390.

32   Henry DH, Abels RI: Recombinant human erythropoietin in the treatment of cancer and chemotherapy-induced anemia: Results of double-blind and open-label follow-up studies. Semin Oncol 1994; 21(suppl):21–28.

33   Bruera E, Fainsinger RL: Clinical management of cachexia and anorexia; in Doyle D, Hanks G, MacDonald N (eds): Oxford Textbook of Palliative Medicine. New York, Oxford University Press, 1993, pp 330–337.

34   Nelson KA, Walsh D, Sheehan FA: The cancer anorexia-cachexia syndrome. J Clin Oncol 1994; 12:213–215.

35   White MC, Kendal-Taylor P: Adrenal hypofunction inpatients taking ketoconazole. Lancet 1985;i:44–45.

36   Moertel C, Schutt AJ, Reitemeier RJ, Hahn RG: Corticosteroid therapy of preterminal gastrointestinal cancer. Cancer 1974;33:1607–1609.

37   Bruera E, Carraro S, Roca E, Barugel M, Chacon R: Action of oral methylprednisolone in terminal cancer patients: A prospective randomized double-blind study. Cancer Treat Rev 1985;69:751–754.

38   Bruera E, MacMillan K, Hanson J, Kuehn N, MacDonald RN: A controlled study of megestrol acetate on appetite, caloric intake, nutritional status and other symptoms in patients with advanced cancer. Cancer 1990;66:1279–1282.

39   MacDonald SM, Hagen N, Bruera E: Proximal muscle weakness in a patient with hepatocellular carcinoma. J Pain Symptom Manage 1994;9:346–350.

40   Bruera E, Carrara S, Roca E, Barugel M, Chacon R: Double-blind evaluation of the effects of mazindol on pain, depression, anxiety, appetite and activity in terminal cancer patient. Cancer Treat Rep 1985;70:295–297.

41   Bruera E, Chadwick S, Brenneis C: Methylphenidate associated with narcotics for the treatment of cancer pain. Cancer Treat Rep 1987;71:120–127.

42   MacVicar MG, Winningham ML: Promoting functional capacity of cancer patients. Cancer Bull 1986;38:235–239.

43   St Pierre BA, Kasper CE, Lindsey AM: Fatigue mechanisms in patients with cancer: Effects of tumor necrosis factor and exercise on skeletal muscle. Oncol Nurs Forum 1992;19:419–425.

44   Brophy LR, Sharp EJ: Physical symptoms of combination biotherapy: A quality of life issue. Oncol Nurs Forum 1991;18(suppl):25–30.

45   Hoskin PJ: Radiotherapy in symptom management; in Doyle D, Hanks G, MacDonald N (eds): Oxford Textbook of Palliative Medicine. New York, Oxford Medical Publications, 1993, pp 117–129.

46   Coia LR: Practice patterns of palliative care for the United States 1984–1985. Int J Radiat Oncol Biol Phys 1988;14:1261–1269.

47   Tong D, Gillick L, Hendrickson FR: The palliation of symptomatic osseous metastases: Final results of the study by the RTOG. Cancer 1982;50:893–899.

48   Blitzer P: Reanalysis of the RTOG study of the palliation of symptomatic osseous metastasis. Cancer 1985;55:1468–1472.

49   Munro AJ, Sebag-Montefiore D: Opportunity cost – A neglected aspect of cancer treatment. Br J Cancer 1992;65:309–310.

50   Jaakkimainen L, Goodwin PJ, Pater J, Warde P, Murray N, Rapp E: Counting the costs of chemotherapy in a National Cancer Institute of Canada randomized trial in non-small cell lung cancer. J Clin Oncol 1990;8:1301–1309.

51   Grilli R, Oxman AD, Julian JA: Chemotherapy for advanced non-small cell lung cancer: How much benefit is enough? J Clin Oncol 1993;11:1866–1872.

52   Chevalier T: Chemotherapy for advanced NSCLC. Will meta-analysis provide the answer? Chest 1996;109(suppl 5):107–109.

53   Marino P, Pampallona S, Preatoni A, Cantoni A, Invernizzi F: Chemotherapy vs. supportive care in advanced non-small cell lung cancer. Results of a meta-analysis of the literature. Chest 1994;106:861–865.

54   Non-Small Cell Jung Cancer Collaborative Group: Chemotherapy in non-small cell lung cancer: A meta-analysis using updated data on individual patients from 52 randomised clinical trials. BMJ 1995;311:899–909.

55   Raby B, Pater J, Mackillop WJ: Does knowledge guide practice? Another look at the management of non-small cell lung cancer. J Clin Oncol 1995;13:1904–1911.

Prof. James F. Cleary, University Hospital and Clinics, Madison, Wisc., USA

Schiller JH (ed): Updates in Advances in Lung Cancer. Prog Respir Res.
Basel, Karger, 1997, vol 29, pp 173–186

Chapter 11
....

The Molecular Biology of Lung Cancer

Tadao Ishida, David P. Carbone

The Vanderbilt Cancer Center, Nashville, Tenn., USA

Introduction

Normal cells respond to both external and internal cues that tell them to divide or to stop dividing. A fundamental property of the cancer cell is its ability to ignore these cues. How do lung cancer cells lose this regulated growth? Lung cancer is thought to develop after a series of morphological and genetic changes that may take years to occur. The precise order of molecular events which lead to invasive cancer (if indeed there is an order) has not yet been established, though this is being actively studied.

Nitrosamines are reactive substances in tobacco smoke thought to play a role in carcinogenesis. Another class of carcinogens in tobacco smoke are the polycyclic aromatic hydrocarbons, such as benzo(a)pyrene. The compound requires metabolic activation to become the ultimate carcinogenic metabolite, benzo(a)pyrene diol epoxide (BPDE). These substances cause covalent carcinogen adducts to form with many molecules within DNA. Recently, the distribution of BPDE adducts among particular codons of the p53 gene in BPDE-treated Heha cells and bronchial epithelial cells was mapped. Strong and selective adduct formation occurred at guanine position in codons 157, 248 and 273. The coincidence of mutational hotspots and adduct hotspots adds strength to the hypothesis that benzo(a)pyrene metabolites or structurally related compounds derived from cigarette smoke are involved in the malignant transformation of human lung cells [1].

Broadly speaking, DNA damage can result in the 'activation' of a growth stimulatory gene (a 'dominant' oncogene) or the inactivation of a growth suppressing gene (a 'tumor suppressor' gene). In the simplest case, activation of a dominant oncogene involves a single nucleotide substitution, resulting in the alteration of an amino acid in the gene product. This would only have to occur in one of the two copies of gene. Sequence analysis of these genes in a variety of tumors has revealed activating point mutations or regulatory alterations.

Tumor suppressor genes can lose function by point mutation as well, but both copies must lose function to result in complete loss of the involved pathway in a particular cell. This second copy can be lost by point mutation or deletion, but hypermethylation is also increasingly being shown as an alternative mode of inactivation.

Tumor Suppressor Genes in Lung Cancer

In contrast to the gain of function characteristic of dominant oncogenes, tumor suppressor genes cause dysregulated growth primarily through loss of function. Allele loss is a hallmark of chromosome regions harboring recessive oncogenes. Perhaps the most common way for a cell to 'uncover' a single defective tumor suppressor gene would be by the deletion of the other, normal, allele. This deletion can be large and span hundreds of other genes, but nonetheless, the detection of such 'hemizygous' loss can be an important indication that a tumor suppressor gene lies somewhere within the limits of the deletion.

Lung cancer has been associated with loss of heterozygosity at several distinct genetic loci including frequent losses on chromosomes 3p, 9p, 13q and 17p. Many of the known tumor suppressor genes (such as Rb, p16 and p53) have been found, perhaps not surprisingly, to be involved in control of the cell cycle (fig. 1). The data describing the involvement of the most important of these genes will be outlined below.

Rb

The prototype tumor suppressor gene is *rb*, a nuclear phosphoprotein involved in cell cycle regulation located on chromosome 13q. Particularly critical for the control of cell growth is the pathway involving the G1 cyclin-dependent kinases that regulate the Rb family of proteins, which in turn control E2F transcription factor activity. The Rb protein is thought to bind and sequester E2F-related transcription factors which promote cell cycling. Phosphorylation of Rb protein by the cyclin-dependent kinases causes release of the E2F which then stimulates cell division. Many tumors, including lung, have acquired abnormalities of Rb. Most SCLC (>95%) have absent or abnormal Rb protein, while only modest fractions of NSCLC are affected [2]. Re-introduction of Rb into lung cancer cell lines suppresses tumorigenicity [3], but there appears to be no correlation of tumor Rb status with patient survival [4].

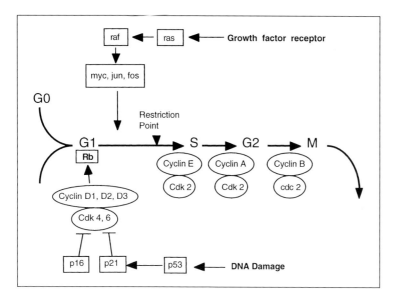

Fig. 1. Cell cycle regulators. Movement through the cycle is positively regulated by the complexes formed between different cyclins and their catalytic partners the cdks. In contrast, cell cycle progression is retarded by negative regulators (⊢) consisting of diverse cdk inhibitors: p16, the product of the CDNK2 gene, which specifically binds to cdk4 and inhibits cyclinD/cdk4 activity; p21, which induced by p53 in response to DNA damage and binds to multiple cdks. The restriction point is in late G1 at which cells become committed to divide irrespective of extracellular growth signals and may be coincident with the cell cycle block imposed by Rb protein.

p16

The cyclin–dependent kinase inhibitor known as p16 (CDK4I, CDKN2, INK4A, MTSI) has been proposed as a tumor suppressor gene located on chromosome 9p21. Homozygous deletion of CDKN2, which encodes the p16, was detected in 18 (23%) of 77 cell lines established from patients with NSCLC, compared with 1 (1%) of 93 cell lines established from patients with SCLC [5]. And at a high frequency (6/9, 67%), CDKN2 was either deleted or mutated in NSCLC cell lines [6]. Interestingly, in NSCLC there exists an inverse correlation between expression of the p16 and the presence of a functional Rb [7]. p16 inhibits cdk4 which phosphorylates Rb, and therefore the absence of p16 allows uncontrolled cdk4 phosphorylation of Rb and transcriptional activation. This may explain the lack of clinical correlations with Rb status in NSCLC, as virtually all NSCLC have this pathway inactivated, and Rb inactivation is only one of the mechanisms for doing so. Mutations in p16 are occasionally observed in freshly resected NSCLC tumors [8, 9]. In contrast to p16, cyclin D1 activates cdk4-mediated phosphorylation of Rb, and many lung cancers overexpress this protein also tending to circumvent the ability of Rb to modulate cell cycling. Overexpression of cyclin D1 has been related to a reduced local relapse rate [10].

p53

p53 is a nuclear phosphoprotein located on chromosome 17p with specific DNA binding and transcriptional activation capabilities. p53 is very frequently abnormal in lung cancer, with mutation in $> 50\%$ of NSCLC and $> 90\%$ of SCLC [11]. Compared to the limited number of substitutions found in activated *ras,* mutations in p53 are of all types: missense, nonsense, splicing, and large deletions. The missense mutations occur throughout the open reading frame, though they are most frequent in the central exons, 5 through 8. Missense mutations are also associated with prolonged protein half-life, and thus increased steady-state protein levels. This can often be detected by immunohistochemistry, and forms the basis for a simple 'screening' immunohistochemical test of clinical material for the presence of p53 mutations, which is only about 2/3 correlated with the presence of a mutation, however [12]. The locations of missense mutations in p53 are similar between different tumor types, but there may be a preference for mutations between codon 150 and 160 in NSCLC, and codon 175 mutations are very common in colon cancer, but are uncommon in lung cancer. The significance of these differences is unclear. Mutations in both synchronous [13] and metachronous [14] lung cancers have been shown to be different, supporting the frequently presumed independent origin of these lesions.

One of the major p53-induced genes known is p21/WAF1, an inhibitor of the cyclin-dependent kinases. Normal p53 can induce p21 which inhibits cdk4, leaving Rb in an unphosphorylated state, inhibiting the cell cycle. In lung cancer, however, p21 is usually expressed at higher than normal levels independent of the status of p53, suggesting a p53-independent pathway for its induction. p21 itself is not frequently mutated [15], also suggesting that it is not the only mediator of p53 action. Expression of p53 is also increased by agents which cause DNA damage, and the increased expression of p53 normally causes cell cycle arrest. Mutants in p53 are unable to mediate this arrest, and it is thought that p53-mediated cycle arrest allows the cell time to repair its DNA before replicating it in S phase. Without this arrest, DNA replication would proceed to 'fix' the mutation in all of the progeny of that cell. Introduction of wild-type p53 into a number of tumors inhibits growth in vitro and in vivo [16].

The presence of mutations in p53 has an uncertain and probably insignificant relationship with clinical outcome. Early studies reported a strong adverse prognostic impact, but subsequent studies have found either no association or even a survival advantage in patients with tumors bearing p53 mutations [12]. There may however be some association with chemosensitivity both in vitro [17] and in vivo [18]. The association of p53 with response to DNA damage, e.g. that induced standard cancer therapeutics, mandates that future studies of its clinical impact involve carefully controlled, uniformly staged and treated cohorts of patients, unlike many of the studies currently in the literature.

Antibodies to p53 have been observed in lung cancer patients and appear to be correlated with the presence of a mutation in the tumor. In SCLC, an association has been observed between the presence of antibodies to any tumor cell protein (not just p53) and improved survival. Antibodies to p53 may precede the development of clinical cancer [19].

Other Suppressor Genes

There are a number of chromosomal regions thought to harbor as yet undiscovered tumor suppressor genes. One of these is chromosome 5p, frequently lost in SCLC and some NSCLC and another is 3p, probably the most frequently deleted chromosomal region in lung cancer [20]. Recent progress has been made in localizing candidate genes [21], and a novel gene, FHIT, has been cloned from 3p14 which demonstrates altered splicing in many types of tumors, including lung [22]. The functional and potential clinical and therapeutic relevance of these findings are under investigation.

Oncogene Lesions in Lung Cancer

Many transforming genes carried by the oncogenic retroviruses (prefixed by v-) have been found to have normal cellular homologs called proto-oncogenes (prefixed by c-). Perhaps the best studied example is the retroviral gene from the Harvey and Kirsten rat sarcoma viruses, whose genes v-H-*ras* and v-K-*ras* have cellular homologs c-H-*ras,* c-K-*ras.* Other examples include *myc, raf, myb, fos* and *jun* and many others.

ras

The *ras* genes code for 21-kD proteins attached to the inner surface of the cytoplasmic membrane via posttranslationally added lipid. They are homologous to G proteins and are thought to mediate signal transduction through raf and mitogen-activated protein (MAP) kinase. G proteins exist in two states, GDP- and GTP-bound. Only the GTP-bound form is effective at mediating a growth response, and there is a dynamic interconversion between the two forms. Growth stimuli (mitogens or growth factors) cause the substitution of GTP for GDP, activating *ras,* and intrinsic GTPase activity (with GTPase activating protein, or GAP) catalyzes the conversion of *ras* back to the inactive form. *Ras* mutations are frequent in human tumors and usually occur by point mutations at codons 12, 13 or 61. Most *ras* mutants are defective in GTPase activity and thus are 'locked' into the growth stimulatory GTP-bound form.

In one series of lung cancer cell lines, Mitsudomi et al. [23] found *ras* mutations exclusively in NSCLC, and predominantly in K-*ras* codon 12. The lack of *ras* mutations in SCLC suggests a molecular specificity for this genetic abnormality and may reflect different pathogenetic mechanisms. In studies of 80 resected lung cancer specimens, Rodenhuis et al. [24] also found mutations exclusively in adenocarcinoma. They found that K-*ras* mutations tended to occur in tumors that were smaller and less differentiated than those without mutations. The presence of a K-*ras* codon 12 mutation was also strong adverse prognostic factor for survival [25], though no correlation with *ras* mutations and chemosensitivity has been observed. Inhibition of *ras* expression with an antisense retrovirus [26] and by specific small molecule inhibitors [27] has shown activity in preclinical models and human clinical trials are ongoing or planned.

myc

In contrast to *ras* genes which are activated by point mutations, *myc* genes are uncommonly mutant, but rather appear to be usually activated by overexpression either by up-regulation or gene amplification. N- and L-*myc* gene amplification is seen in 10–20% of SCLC tumor samples taken prior to chemotherapy [28]. In cell lines established after chemotherapy and clinical relapse, 11 of 25 had amplified *myc* genes (5 c-*myc,* 3 N-*myc* and 3 L-*myc*) (44%) [29]. C-*myc* amplification is associated with the variant phenotype of SCLC and with shortened patient survival [29]. When RNA expression is evaluated rather than DNA amplification, 80–90% of SCLC show overexpression relative to normal or fetal lung tissues [30]. NSCLC only very infrequently shows amplification of c-*myc* (2/47), and these few appear to be adenocarcinomas with normal *ras* genes [31]. Thus, *myc* amplification is not only associated with class of lung tumor (SCLC), but also with prior exposure to chemotherapy and a particular subtype (variant SCLC), as well as poor survival. Inhibition of *myc* gene expression with an antisense construct also inhibits lung cancer cell growth [32].

HER2/neu *and EGF Receptors*

Tyrosine kinase growth factor receptors may play a role in the constitutive signaling of tumor cells to grow without appropriate controls. *c-erb-B1* encodes the epidermal growth factor (EGF) receptor (EGFr), and *c-erb-B2* (also called

HER-2/*neu*) encodes a gene originally isolated from rat neuroblastomas by virtue of structural similarity to EGFr. In the rat neuroblastoma, a mutation in the transmembrane domain of *neu* resulted in constitutive receptor activation. Human studies have identified no mutations but rather amplification and overexpression of HER-2/*neu* with consequent dysregulated signaling. Recently, an anti-HER-2 monoclonal antibody that induced apoptosis was reported [33]. An exciting new approach to killing cells expressing HER2/*neu* involves the delivery of a gene encoding for an intracellular single chain antibody. This is a recombinant molecule designed to bind to the receptor in the endoplasmic reticulum and keep it from getting to the cell surface, and displays potent and specific cytotoxicity [34].

Gene amplification of the *c-erbB-2* is rare in NSCLC, while overexpression is relatively common in NSCLC and uncommon in SCLC [35]. The highest expression levels are seen in adenocarcinomas. Immunohistochemical HER2/*neu* overexpression is also correlated with decreased survival in NSCLC [36] and to chemoresistance [37]. EGF receptors are found predominantly on NSCLC as well, and the production of EGF-like activity by lung cancer cells raises the possibility of an autocrine loop. Expression of EGF receptors also has been shown to be an adverse prognostic factor [38].

Growth Factors and Autocrine Growth Loops

An autocrine growth loop is said to exist if a tumor cell expresses both a growth factor and its receptor such that a self-stimulatory growth loop is observed. Autocrine systems are present in normal cells but these are tightly regulated to respond only to physiological signals and counter-regulatory systems coexist to maintain a balanced growth pattern. This balance is disrupted in cancer cells.

GRP. One of the best characterized autocrine growth loops in SCLC involves GRP, which is the 28 amino acid long mammalian homolog of the amphibian peptide bombesin. GRP is physiologically expressed in neuroendocrine cells in embryonic lung tissues, the central nervous system and the gut-endocrine system. GRP is secreted by most SCLC and 10–20% NSCLC [39]. Specific high-affinity receptors for GRP are also expressed in SCLC tumor specimens and cell lines [40]. The GRP receptor is a member of the guanine nucleotide binding protein-coupled (so-called G protein) receptor superfamily with seven predicted hydrophobic transmembrane domains. A monoclonal antibody specific for the carboxyl-terminus of GRP has been shown to inhibit in vivo growth of SCLC cell lines as well as tumor xenografts in nude mice [41], presumably through the inhibition of GRP peptide antagonists. A phase I-II clinical trial evaluating the GRP-receptor blocking approach in SCLC patients has recently been completed, but not yet published.

Insulin-Like Growth Factor (IGF). The role of the IGF-1 receptor (IGF-1r) in cancer cells has been investigated intensively. In certain systems, the IGF-1r appears to be essential for malignant transformation. Fetal fibroblasts with a disruption of the IGF-1r gene cannot be transformed by SV40 T antigen. The IGFs and their receptors are very important in the lung development and growth of cells in respiratory system as well. In many lung cancer lines, IGF-1 and IGF-1r mediate autocrine proliferation [42] that can be blocked by monoclonal antibodies to IGF-1 or its receptor and by somatostatin analogs [43, 44]. A recent study has shown that treatment of human lung cancer cells with a recombinant adenovirus expressing antisense IGF-1r reduces receptor expression, inhibits colony formation in soft agar, and improves the survival of nude mice with pre-existing intraperitoneal tumors [45]. And a soluble IGF-1r can induce massive apoptosis of tumor cells in vivo [46]. Expression of IGF-II has recently been reported as a negative prognostic factor [47].

Other Oncogenes

A variety of other oncogenes have been studied in lung cancer. C-*myb* is a nuclear oncogene whose level has been inversely associated with differentiation. It is expressed in SCLC but not NSCLC [48]. c-*raf* is an oncogene in the *ras* signal transduction pathway that is located at 3p25 and is frequently deleted in lung cancer, including every case of SCLC examined [49]. There is evidence that c-kit, a tyrosine kinase receptor thought to be involved in hematopoiesis, and its ligand, stem cell factor (SCF), are simultaneously expressed in SCLC, but not NSCLC [50], forming an autocrine growth stimulatory loop [51]. Similarly, platelet-derived growth factor (PDGF), a potent mitogen, and its receptor are both found to be expressed in primary lung cancers and cell lines[52]. Both of these also form potential autocrine growth self-stimulatory loops. The proto-oncogene bcl-2 encodes a protein that inhibits programmed cell death (apoptosis), and decreased apoptosis is a mechanism proposed for dysregulated growth in tumors. Bcl-2 protein is detectable in most SCLC [53] and about 25% of NSCLC. Mutations in the molecular target for etoposide, topoisomerase II, have been identified in tumors after treatment [54], and these mutations mediate resistance to the chemotherapeutic.

Angiogenesis

Successful tumors need a blood supply to grow beyond microscopic size and have subverted a variety of normal cellular pathways to achieve this end. A number of factors, including thrombospondin and vascular endothelial growth factor produced by tumors, act upon defined cell surface receptors in capillary endothelia to promote microvessel ingrowth into solid tumors. Tumors with high microvessel density have an adverse prognosis in stage I NSCLC [55] and a higher incidence of metastases [56].

Angiostatin is an angiogenesis inhibitor that selectively instructs the endothelium to become refractory to angiogenic stimuli. Inhibitors of angiogenesis cause dramatic inhibition of metastases and tumor growth in animal model systems [57] and clinical trials using antiangiogenic natural substances or blocking antibodies are planned or underway. These agents have significant potential usefulness in the treatment of all forms of cancer including lung cancer.

Telomeres in Lung Cancer

The ends of all eukaryotic chromosomes consist of tandem repeats of simple DNA sequences and are called telomeres. These sequences are very important for the stability and fidelity of chromosome replication, and cancer cells often develop ring chromosomes or translocations involving these regions. Recent information appears to indicate that these sequences are involved in cellular senescence as well. Germ cells have longer telomeres (a greater number of repeats) than terminally differentiated cells, and as cells 'senesce' or terminally differentiate, these telomeres shorten. Germ cells, and most cancer cells, have an enzyme called telomerase which keeps the telomeres long, and presumably assists in maintaining the ability to divide indefinitely. This has been observed in lung cancer cells [58] and associated with mutations in Rb [59]. Inhibition of telomerase and consequent induction of senescence is an attractive new approach to cancer therapy.

Oncoproteins as Immunologic Targets

Lung cancer is not an easy target for immunotherapeutic approaches for a number of reasons including lack of clearly identified target antigens and possibly poor immunogenicity. However, the numerous molecularly characterized genetic lesions involved in the development of lung cancer could result in the production of proteins which would be very attractive targets for immunotherapy, if the mutant form could be recognized as distinct from the normal form. This recognition would then be absolutely specific for tumor and not normal tissue. The finding that intracellular proteins such as p53 are normally processed and presented on the surface of the cell in the context of class I MHC molecules, makes this approach for mutant intracellular proteins possible [60]. Therapeutic responses have been seen in animal model tumors by immunotargeting of mutations in p53 [61–63]. It is possible that an effective combined modality therapy which incorporates immunotherapy could be developed in the near future.

Conclusions

Recent advances in the molecular biology of lung cancer may allow the development and application of specific therapeutics which may be not only more effective but less toxic to normal tissues. For example, the function and expression of oncogenes can be directly modulated. Wild-type p53 can be introduced into local tumor deposits with observable effects [16], and the function of mutant p53 modulated by small molecules or antibodies. Interference with growth factor stimulation using gene therapeutic approaches [45] or antibodies [34] or small peptides [64] holds promise for modulation of cancer growth. Interference with the lipid modification of ras is another such approach. The ability to inhibit E2F [65] may be particularly effective in impeding a wide variety of proliferative events because the control of E2F activity appears to the end result of G1 regulatory cascades.

Lung cancer has long been recognized as being a heterogeneous collection of diseases with different histologies, growth properties, endocrine and clinical properties. Molecular biology has revealed a multitude of genetic lesions which have given rise to new levels of complexity, but there are certain patterns emerging from these data. We can now relate histological types of lung cancers and even clinically relevant subsets within a histologic type to certain patterns of molecular lesions. Loss of DNA material from the short arm of chromosome 3 and p53 mutations have been detected in carcinoma in situ lesion and 3p allele loss has been observed in preneoplastic lesion. Loss of chromosomal 9p, bcl-2 overexpression, and microsatellite instability are common in preneoplasia as well. These preneoplastic lesions may be the best targets for some of the novel therapeutic approaches.

References

1   Denissenko MF, Pao A, Tang M-S, Pfeifer GP: Preferential formation of benzo[a]pyren adducts at lung cancer mutational hotspots in P53. Science 1996;274:430–432.
2   Harbour JW, Lai S-L, Whang-Peng J, Gazdar AF, Minna JD, Kaye FJ: Abnormalities in structure and expression of the human retinoblastoma gene in SCLC. Science 1988;241:353–357.
3   Kratzke RA, Shimizu E, Geradts J, Gerster JL, Segal S, Otterson GA, Kaye FJ: RB-mediated tumor suppression of a lung cancer cell line is abrogated by an extract enriched in extracellular matrix. Cell Growth Differ 1993;4:629–635.

4   Shimizu E, Coxon A, Otterson GA, Steinberg SM, Kratzke RA, Kim YW, Fedorko J, Oie H, Johnson BE, Mulshine JL, et al: RB protein status and clinical correlation from 171 cell lines representing lung cancer, extrapulmonary small cell carcinoma, and mesothelioma. Oncogene 1994;9:2441–2448.

5   Kelley MJ, Nakagawa K, Steinberg SM, Mulshine JL, Kamb A, Johnson BE: Differential inactivation of CDKN2 and Rb protein in non-small-cell and small-cell lung cancer cell lines. J Nat Cancer Inst 1995;87:756–761.

6   De Vos S, Miller CW, Takeuchi S, Gombart AF, Cho SK, Koeffler HP: Alterations of CDKN2 (p16) in non-small cell lung cancer. Genes Chromosom Cancer 1995;14:164–170.

7   Otterson GA, Khleif SN, Chen W, Coxon AB, Kaye FJ: CDKN2 gene silencing in lung cancer by DNA hypermethylation and kinetics of p16INK4 protein induction by 5-aza-2′-deoxycytidine. Oncogene 1995;11:1211–1216.

8   Rusin MR, Okamoto A, Chorazy M, Czyzewski K, Harasim J, Spillare EA, Hagiwara K, Hussain SP, Xiong Y, Demetrick DJ, Harris CC: Intragenic mutations of the p16(INK4), p15(INK4B) and p18 genes in primary non-small-cell lung cancers. Int J Cancer 1996;65:734–739.

9   Shapiro GI, Edwards CD, Kobzik L, Godleski J, Richards W, Sugarbaker DJ, Rollins BJ: Reciprocal Rb inactivation and p16INK4 expression in primary lung cancers and cell lines. Cancer Res 1995;55:505–509.

10 Betticher DC, Heighway J, Hasleton PS, Altermatt HJ, Ryder WD, Cerny T, Thatcher N: Prognostic significance of CCND1 (cyclin D1) overexpression in primary resected non-small-cell lung cancer. Br J Cancer 1996;73:294–300.

11 Chiba I, Takahashi T, Nau MM, D'Amico D, Curiel D, Mitsudomi T, Buchhagen D, Carbone D, Koga H, Reissmann PT, Slamon DJ, Holmes EC, Minna JD: Mutations in the p53 gene are frequent in primary, resected non-small cell lung cancer. Oncogene 1990;5:1603–1610.

12 Carbone DP, Mitsudomi T, Chiba I, Piantadosi S, Rusch V, Nowak JA, McIntire D, Slamon D, Gazdar A, Minna J: p53 immunostaining positivity is associated with reduced survival and is imperfectly correlated with gene mutations in resected non-small cell lung cancer. A preliminary report of LCSG 871. Chest 1994;106:377S–381S.

13 Sozzi G, Miozzo M, Pastorino U, Pilotti S, Donghi R, Giarola M, De Gregorio L, Manenti G, Radice P, Minoletti F, et al: Genetic evidence for an independent origin of multiple preneoplastic and neoplastic lung lesions. Cancer Res 1995;55:135–140.

14 Yang HK, Linnoila RI, Conrad NK, Krasna MJ, Aisner SC, Johnson BE, Kelley MJ: TP53 and RAS mutations in metachronous tumors from patients with cancer of the upper aerodigestive tract. Int J Cancer 1995;64:229–233.

15 Shimizu T, Miwa W, Nakamori S, Ishikawa O, Konishi Y, Sekiya T: Absence of a mutation of the p21/WAF1 gene in human lung and pancreatic cancers. Jpn J Cancer Res 1996;87:275–278.

16 Roth JA, Nguyen D, Lawrence DD, Kemp BL, Carrasco CH, Ferson DZ, Hong WK, Komaki R, Lee JJ, Nesbitt JC, Pisters KMW, Putnam JB, Schea R, Shin DM, Walsh GL, Dolormente MM, Han C-I, Martin FD, Yen N, Xu K, Stephens et al: Retrovirus-mediated wild-type p53 gene transfer to tumors of patients with lung cancer. Nat Med 1996;2:985–991.

17 Fujiwara T, Grimm EA, Mukhopadhyay T, Zhang WW, Owen-Schaub LB, Roth JA: Induction of chemosensitivity in human lung cancer cells in vivo by adenovirus-mediated transfer of the wild-type p53 gene. Cancer Res 1994;54:2287–2291.

18 Rusch V, Klimstra D, Venkatraman E, Oliver J, Martini N, Gralla R, Kris M, Dmitrovsky E: Aberrant p53 expression predicts clinical resistance to cisplatin-based chemotherapy in locally advanced non-small cell lung cancer. Cancer Res 1995;55:5038–5042.

19 Lubin R, Zalcman G, Bouchet L, Tredanel J, Legros Y, Cazals D, Hirsch A, Soussi T: Serum p53 antibodies as early markers of lung cancer. Nat Med 1995;1:701–702.

20 Whang-Peng J, Kao-Shan C, Lee E, Bunn P Jr, Carney D, Gazdar A, Minna J: A specific chromosome defect associated with human small-cell lung cancer: Deletion 3p$_{(14–23)}$. Science 1982;215:181–182.

21 Sekido Y, Bader S, Latif F, Chen JY, Duh FM, Wei MH, Albanesi JP, Lee CC, Lerman MI, Minna JD: Human semaphorins A(V) and IV reside in the 3p21.3 small cell lung cancer deletion region and demonstrate distinct expression patterns. Proc Nat Acad Sci USA 1996;93:4120–4125.

22 Sozzi G, Veronese ML, Negrini M, Baffa R, Cotticelli MG, Inoue H, Tornielli S, Pilotti S, De Gregorio L, Pastorino U, Pierotti MA, Ohta M, Huebner K, Croce CM: The FHIT gene 3p14.2 is abnormal in lung cancer. Cell 1996;85:17–26.

23   Mitsudomi T, Viallet J, Mulshine JL, Linnoila RI, Minna JD, Gazdar AF: Mutations of *ras* genes distinguish a subset of non-small-cell lung cancer cell lines from small-cell lung cancer cell lines. Oncogene 1991;6:1353–1362.

24   Rodenhuis S, Slebos RJC, Boot AJM, Evers SG, Mooi WJ, Wagenaar SS, van Bodegom PC, Bos JL: Incidence and possible clinical significance of *K-ras* oncogene activation in adenocarcinoma of the human lung. Cancer Res 1988;48:5738–5741.

25   Slebos RJ, Kibbelaar RE, Dalesio O, Kooistra A, Stam J, Meijer CJ, Wagenaar SS, Vanderschueren RG, van Zandwijk N, Moot WJ, Bos JL, Rodenhuis S: *K-ras* oncogene activation as a prognostic marker in adenocarcinoma of the lung. N Engl J Med 1990;323:561–565.

26   Georges RN, Mukhopadhyay T, Zhang Y, Yen N, Roth JA: Prevention of orthotopic human lung cancer growth by intratracheal instillation of a retroviral antisense *K-ras* construct. Cancer Res 1993;53:1743–1746.

27   James GL, Goldstein JL, Brown MS, Rawson TE, Somers TC, McDowell RS, Crowley CW, Lucas BK, Levinson AD, Marsters JJ: Benzodiazepine peptidomimetics: Potent inhibitors of Ras farnesylation in animal cells (see comments). Science 1993;260:1937–1942.

28   Johnson B, Makuch R, Simmons A, Gazdar A, Burch D, Cashell A: *myc* family DNA amplification in small cell lung cancer patients' tumors and corresponding cell lines. Cancer Res 1988;48:5163–5166.

29   Johnson BE, Ihde DC, Makuch RW, Gazdar AF, Carney DN, Oie H, Russell E, Nau MM, Minna JD: *myc* family oncogene amplification in tumor cell lines established from small cell lung cancer patients and its relationship to clinical status and course. J Clin Invest 1987;79:1629–1634.

30   Takahashi T, Obata Y, Sekido Y, Hida T, Ueda R, Watanabe H, Ariyoshi Y, Sugiura T, Takahashi T: Expression and amplification of *myc* gene family in small cell lung cancer and its relation to biological characteristics. Cancer Res 1989;49:2683–2688.

31   Slebos R, Evers S, Wagenaar S, Rodenhuis S: Cellular protooncogenes are infrequently amplified in untreated non-small cell lung cancer. Br J Cancer 1989;59:76–80.

32   Dosaka-Akita H, Akie K, Hiroumi H, Kinoshita I, Kawakami Y, Murakami A: Inhibition of proliferation by L-myc antisense DNA for the translational initiation site in human small cell lung cancer. Cancer Res 1995;55:1559–1564.

33   Kita Y, Tseng J, Horan T, Wen J, Philo J, Chang D, Ratzkin B, Pacifici R, Brankow D, Hu S, Lou Y, Wen D, Arakawa T, Nicolson M: ErbB receptor activation, cell morphology changes, and apoptosis induced by anti-Her2 monoclonal antibodies. Biochem Biophys Res Commun 1996;226:59–69.

34   Deshane J, Siegal GP, Alvarez RD, Wang MH, Feng M, Cabrera G, Liu T, Kay M, Curiel DT: Targeted tumor killing via an intracellular antibody against erbB-2 J Clin Invest 1995;96:2980–2989.

35   Weiner DB, Nordberg J, Robinson R, Nowell PC, Gazdar A, Greene MI, Williams WV, Cohen JA, Kern JA: Expression of the neu gene-encoded protein (P185neu) in human non-small cell carcinomas of the lung. Cancer Res 1990;50:421–425.

36   Kern JA, Schwartz DA, Nordberg JE, Weiner DB, Greene MI, Torney L, Robinson RA: p185neu expression in human lung adenocarcinomas predicts shortened survival. Cancer Res 1990;50:5184–5187.

37   Tsai CM, Chang KT, Wu LH, Chen JY, Gazdar AF, Mitsudomi T, Chen MH, Perng RP: Correlations between intrinsic chemoresistance and HER–2/neu gene expression, p53 gene mutations, and cell proliferation characteristics in non-small cell lung cancer cell lines. Cancer Res 1996;56:206–209.

38   Fontanini G, Vignati S, Bigini D, Mussi A, Lucchi H, Angeletti CA, Pingitore R, Pepe S, Basolo F, Bevilacqua G: Epidermal growth factor receptor expression in non-small cell lung carcinomas correlates with metastatic involvement of hilar and mediastinal lymph nodes in the squamous subtype. European Journal of Cancer 1995;31A:178–183.

39   Moody T, Pert C, Gazdar A, Carney DN, Minna J: High levels of intracellular bombesin characterize human small-cell lung cancer. Science 1981;214:1246–1248.

40   Moody TW, Carney DN, Cuttitta F, Quattrocchi K, Minna JD: High affinity receptors for bombesin/GRP-like peptides on human small cell lung cancer. Life Sci 1985;37:105–113.

41   Cuttitta F, Carney DN, Mulshine J, Moody TW, Fedorko J, Fischler A, Minna JD: Bombesin-like peptides can function as autocrine growth factors in human small-cell lung cancer. Nature 1985;316:823–826.

42 Nakanishi Y, Mulshine JL, Kasprzyk PG, Natale RB, Maneckjee R, Avis I, Treston AM, Gazdar, AF, Minna JD, Cuttitta F: Insulin-like growth factor-I can mediate autocrine proliferation of human small cell lung cancer cell lines in vitro. J Clin Invest 1988;82:354–359.

43 Macaulay VM, Everard MJ, Teale JD, Trott PA, Van Wyk JJ, Smith IE, Millar JL: Autocrine function for insulin-like growth factor I in human small cell lung lines and fresh tumor cells. Cancer Res 1990;50:2511–2517.

44 Taylor JE, Bogden AE, Moreau JP, Coy DH: In vitro and in vivo inhibition of human small cell lung carcinoma (NCI-H69) growth by a somatostatin analogue. Biochem Biophys Res Commun 1988;153:81–86.

45 Lee C-T, Wu S, Gabrilovich D, Chen H, Nadaf-Rahrov S, Ciernik IF, Carbone DP: Antitumor effects of an adenovirus expressing antisense insulin-like growth factor I receptor on human lung cancer cell lines. Cancer Res 1996;56:3038–3041.

46 D'Ambrosio C, Ferber A, Resnicoff M, Baserga R: A soluble insulin-like growth factor I receptor that induces apoptosis of tumor cells in vivo and inhibits tumorigenesis cancer. Res 1996;56: 4013–4020.

47 Takanami I, Imamuma T, Hashizume T, Kikuchi K, Yamamoto Y, Yamamoto Y, Kodaira S: Insulin-like growth factor-II as a prognostic factor in pulmonary adenocarcinoma. J Surg Oncol 1996;61: 205–208.

48 Griffin C, Baylin S: Expression of the c-*myb* oncogene in human small cell lung carcinoma. Cancer Res 1985;45:272–275.

49 Graziano SL, Pfeifer AM, Testa JR, Mark GE, Johnson BE, Hallinan EJ, Pettengill OS, Sorenson GD, Tatum AH, Brauch H, et al: Involvement of the RAF1 locus, at band 3p25, in the 3p deletion of small-cell lung cancer. Genes Chromosom Cancer 1991;3:283–293.

50 Sekido Y, Obata Y, Ueda R, Hida T, Suyama M, Shimokata K, Ariyoshi Y, Takahashi T: Preferential expression of c-kit protooncogene transcripts in small cell lung cancer. Cancer Res 1991;51:2416–2419.

51 Krystal GW, Hines SJ, Organ CP: Autocrine growth of small cell lung cancer mediated by coexpression of c-kit and stem cell factor. Cancer Res 1996;56:370–376.

52 Antoniades HN, Galanopoulos T, Neville GJ, O'Hara CJ: Malignant epithelial cells in primary human lung carcinomas coexpress in vivo platelet-derived growth factor (PDGF) and PDGF receptor mRNAs and their protein products. Proc Natl Acad Sci USA 1992;89:3942–3946.

53 Yan JJ, Chen FF, Tsai YC, Jin YT: Immunohistochemical detection of Bcl-2 protein in small cell carcinomas. Oncology 1996;53:6–11.

54 Kubo A, Yoshikawa A, Hirashima T, Masuda N, Takada M, Takahara J, Fukuoka M, Nakagawa K: Point mutations of the topoisomerase II alpha gene in patients with small cell lung cancer treated with etoposide. Cancer Res 1996;56:1232–1236.

55 Harpole DH Jr, Richards WG, Herndon JE, Sugarbaker DJ: Angiogenesis and molecular biologic substaging in patients with stage I non-small cell lung cancer. Ann Thorac Surg 1996;61:1470–1476.

56 Yuan A, Yang PC, Yu CJ, Lee YC, Yao YT, Chen CL, Lee LN, Kuo SH, Luh KT: Tumor angiogenesis correlates with histologic type and metastasis in non-small-cell lung cancer. Am J Respir Crit Care Med 1995;152:2157–2162.

57 O'Reilly MS, Holmgren L, Chen C, Folkman J: Angiostatin induces and sustains dormancy of human primary tumors in mice. Nat Med 1996;2:689–692.

58 Shirotani Y, Hiyama K, Ishioka S, Inyaku K, Awaya Y, Yonehara S, Yoshida Y, Inai K, Hiyama E, Hasegawa K, et al: Alteration in length of telomeric repeats in lung cancer. Lung Cancer 1994; 11:29–41.

59 Hiyama K, Ishioka S, Shirotani Y, Inai K, Hiyama E, Murakami I, Isobe T, Inamizu T, Yamakido M: Alterations in telomeric repeat length in lung cancer are associated with loss of heterozygosity in p53 and Rb. Oncogene 1995;10:937–944.

60 Hunt DF, Henderson RA, Shabanowitz J, Sakaguchi K, Michel H, Sevilir N, Cox AL, Appella E, Engelhard VH: Characterization of peptides bound to the class I MHC molecule HLA-A2.1 by mass spectrometry. Science 1992;255:1261–1263.

61 Noguchi Y, Richards EC, Chen YT, Old LJ: Influence of interleukin 12 on p53 peptide vaccination against established Meth A sarcoma. Proc Natl Acad Sci USA 1995;92:2219–2223.

62   Gabrilovich DI, Ciernik IF, Carbone DP: Dendritic cells in antitumor immune responses. I. Defective antigen presentation in tumor-bearing hosts. Cell Immunol 1996;170:101–110.

63   Gabrilovich DI, Nadaf S, Corak J, Berzofsky JA, Carbone DP: Dendritic cells in anti-tumor immune responses. II. Dendritic cells grown from bone marrow precursors, but not mature DC from tumor-bearing mice are effective antigen carriers in the therapy of established tumors. Cell Immunol 1996;170:111–119.

64   Moody TW, Venugopal R, Zia F, Patierno S, Leban JJ, McDermed J: BW2258U89: A GRP receptor antagonist which inhibits small cell lung cancer growth. Life Sci 1995;56:521–529.

65   Ishizaki J, Nevis J, Sullenger BA: Inhibition of cell proliferation by an RNA ligand that selectively blocks E2F function. Nat Med 1996;2:1386–1389.

David Carbone, MD, 1161 21st Avenue, South,
1956 The Vanderbilt Clinic, Nashville, TN 37232-5536 (USA)
Tel. (615) 936-3321, Fax (615) 936-3322, E-mail carbondp@ctrvax.vanderbilt.edu

Subject Index

Cisplatin, *see also* CAP chemotherapy, MVP chemotherapy
  chemoradiotherapy trials in locally advanced non-small cell lung cancer 38–42, 45–47
  docetaxel combination therapy 110, 111
  gemcitabine combination therapy in non-small cell lung cancer 98–101
  non-small cell lung cancer trials with surgery 25–27
  stage IV non-small cell lung cancer treatment 57–59, 66–69
  toxicity 58, 59, 68
C-*myb*, mutations in lung cancer 180
Cough, *see* Dyspnea
C-*raf*, mutations in lung cancer 180
Cyclophosphamide, *see* CAP chemotherapy
Cytokines, non-small cell lung cancer trials 31

Difluorodeoxycytidine, *see* Gemcitabine
Docetaxel
  mechanism 106
  non-small cell lung cancer trials
    combination therapy
      cisplatin 110, 111
      vinblastine 112
      vinorelbine 112
    monotherapy
      phase I trials 107, 108
      phase II trials 108–110
    radiation sensitization 113
  small cell lung cancer treatment 142
  stage IV non-small cell lung cancer treatment 64, 65
  tumor specificity 106
Doxorubicin, *see* CAP chemotherapy
Dyspnea
  assessment 156
  incidence in lung cancer 150, 163
  opioid therapy 163, 164
  oxygen therapy 163
  respiratory sedatives in therapy 164

Edmonton Symptom Assessment System (ESAS) 157

Epidermal growth factor receptor, gene mutations in lung cancer 178, 179
ESAS, *see* Edmonton Symptom Assessment System
Etoposide
  chemoradiotherapy trials in locally advanced non-small cell lung cancer 38, 41, 45, 46, 48
  mechanism 143
  non-small cell lung cancer trials with surgery 28, 29
  palliative treatment 168
  small cell lung cancer treatment
    chronic oral administration 143
    cisplatin combination therapy 144
    elderly patients 144, 145
Etretinate, cancer prevention 7
Exercise, cancer patients 166

Fatigue
  assessment 157
  management
    anemia correction 164
    drug interventions 165, 166
    exercise 166
    nutrition 165

Gemcitabine
  mechanism 65, 104, 142
  non-small cell lung cancer treatment
    clinical symptom release 95–97
    combination therapy
      carboplatin 101
      cisplatin 98–101
      ifosfamide 102, 103
    dose response 104
    monotherapy 91, 92, 94, 95, 103
  small cell lung cancer treatment 142, 143
  stage IV non-small cell lung cancer treatment 65
  tumor specificity 81
GRP, autocrine growth loops in lung cancer 179

HER2/*neu*, expression in lung cancers 179